CW00759621

To

Air Chief Marshal Sir Stephen Dalton

with compliments & Best Wishes

07 March 2013 .

THE PURPLE LEGACY

*Indian Air Force Helicopters in
Service of the Nation*

THE PURPLE LEGACY

Indian Air Force Helicopters in Service of the Nation

Air Commodore Rajesh Isser VM

PENTAGON PRESS

THE PURPLE LEGACY
Indian Air Force Helicopters in Service of the Nation
by *Air Commodore Rajesh Isser VM*

ISBN 978-81-8274-688-6

First Published in 2012

Copyright © Air Commodore Rajesh Isser VM

All rights reserved. No part of this publication may be reproduced, stored in a retrieval system, or transmitted, in any form or by any means, electronic, mechanical, photocopying, recording, or otherwise, without first obtaining written permission of the copyright owner and the publishers.

The views and opinion expressed in the book are the individual assertion and opinion of the Author and the Publisher do not take any responsibility for the same in any manner whatsoever. The same shall solely be the responsibility of the Author.

Published by
PENTAGON PRESS
206, Peacock Lane, Shahpur Jat,
New Delhi-110049
Phones: 011-64706243, 26491568
Telefax: 011-26490600
email: rajan@pentagonpress.in
website: www.pentagonpress.in

Printed at Syndicate Binders, Noida.

To Teesta & Pratyush

*Representing a new generation that must acknowledge
the air warrior of yesteryears who enabled our country
to grow and flourish, secure in the knowledge
that it was ably protected.*

(Purple is the Colour on mixing IAF Blue,
Indian Army Olive Green and Navy Blue)

Contents

Acknowledgements

Foremost, I thank the Chief of the Air Staff for approving and encouraging this project.

Air Mshl Daljit Singh, DG Air (Ops) has always been a pillar of encouragement for such endeavours and I thank him for his support. My gratitude to AVM PN Pradhan, ACAS Ops (T&H) for giving me a free hand to carry out research away from our busy schedule. I wish to thank AVM Arjun Subramaniam, ACAS Ops (Space) who provided the crucial guidance during the process of preparing this book.

My gratitude to Sqn Ldr RTS Chinna (retd) of the Centre for Armed Forces Historical Research for the invaluable discussion, debates and research material that he was so forthcoming with. I wish to also thank various contributors of personal articles including Gp Capt BS Bakshi (retd), Air Cmde CM Singla (retd), Air Cmde Melville Rego (retd), Air Cmde HS Ahluwalia (retd), AVM M Bahadur, Air Cmde NM Samuel and many others. I also acknowledge the great service being provided to the Defence Services by the Bharat Rakshak website.

The exhaustive data and personal accounts could not have been gathered in such a short time without the whole-hearted support of all helicopter units (HUs) and their Commanding Officers. I was immensely benefitted by the personal interaction during my visits to all HUs, and was completely overwhelmed by the enthusiasm, josh and of course the warm hospitality. My best wishes to all for continuing the good work. I do acknowledge the assistance to me in the project by Sgts SB Valli, RK Mishra and KK Verma.

All views are personal and not necessarily of the IAF or any other institution. Any factual or other errors are attributable to me alone.

Prologue

There are hardly any books by IAF helicopter pilots to document the rich experience of the fleet– in fact there are none. This is indeed sad since some invaluable experience, tales and lore may have been lost forever. Which brings me to the main reason for this book – to encourage units, squadrons and even retired officers to put pen to paper and record the legacy that the nation has every reason to be proud of. The other reason is to very briefly sketch this legacy for future helicopter pilots to be inspired and motivated to carry on the good but difficult work. In the hustle and bustle of a very technological-intensive service life, sometimes we do not have time for history, and they say 'ignore history at your own peril'. However, this book does not pretend to be a wholesome historical chronicle of IAF helicopters. It aims to act as catalyst for more readings and a lot more research. Even by moderate standards, each unit's history would require a separate volume.

It was a humbling experience going through all the accounts, data and citations. Pilots who had just converted went into war armed with great self-belief and a commitment to the Indian Jawan. Shot down aircrew escaping back through tortuous terrain to fly and fight again. Maintenance engineers during a conflict risking all to salvage and repair helicopters to make them airworthy again. Aircrew flying daily missions over many years over inhospitable terrain, knowing full well that a thin line separated them from death – and the list goes on. For every one rewarded, there are many others whose stories have not been told, yet may they rest or sleep in peace for they did their duty – and all holy scriptures tell us that it is the only thing that matters in the ultimate analysis.

Some tales bring out the peculiarities of all that goes into making a helicopter pilot. The pressure to mature fast, independent decision making as young officers, empathy for the ground troops, coordinating with civilians, NGOs, humanitarian actors etc; dealing with constantly new variables even as the previous one's are not behaving as per plan, extending the envelopes of performances of man and machine and so on. It is no wonder that internationally, in the UN and other forums, an IAF helicopter pilot has the highest attention and respect from peers. Our record in avoiding collateral damage evokes envy and queries from all. Perhaps it is to do with the mature and free-thinking ethos of our service.

Lately, a number of stories in the media have appeared to downplay or even denigrate IAF helicopters and in particular their commitment to jointness in war and peace. I owe it to the martyrs and air warriors of yesteryears to set the record straight. Indeed, the authors of JAAII-86 which delineated specific helicopter responsibilities and capability-building to each service, may have never imagined on how successful and economical it would prove for the country a quarter-century later. Superlative performances in all past wars and conflicts should give all ground forces confidence in the IAF; the tremendous mobilisation enabled by the IAF's medium lift helicopter fleet during the Chinese faceoff in 1986-87 (Op Falcon and Op Chequerboard) in the Northeast being a case in point. With the ongoing modernization and growth of the fleet in the immediate future, the nation can take for granted better capabilities and the same, if not better, commitment from IAF air warriors. Each unit has a large collection of mementoes, letters of appreciations, citations from every agency in India and abroad. It was not possible to include the whole or even a sizeable lot. Only some selected ones have been chosen to make essential points where required. Some of the diary entries and personal accounts have not been touched to retain the flavour of the times when they were written.

The book starts with a look at the initial years of the first unit formation and transformation over the decades. Chapter II scans the horizon around us on how armed forces have been using helicopters and structures of command and control that are evolving. The issue of true integration comes out quite clear. Chapter III to V bring out the heroic exploits of IAF helicopters in the 1962, 1965 and 1971 wars. It will be evident that where dynamic ground commanders, such as Gen Sagat Singh in 1971, have understood and respected air power and the professional advice of airmen, Indian armed forces have performed beyond expectations. Chapter VI on Op Pawan in Sri Lanka highlights how jointness was utilized with great effect by the IPKF. While the overall success of India's foray into Sri Lanka in 1987 is a matter of debate, the joint air-land campaign of the IPKF is a role-model that needs deeper study for its sheer effort.

Chapter VII and VIII chronicle the Herculean effort of IAF helicopters in nation-building in two critical regions– J&K and the North East. But, in no way that meant the Indian Army had been neglected. The next chapter (IX) too amplifies the huge commitment to the army jawan over the decades. It also brings out the difficulty in balancing this act with the never-ending demands of the nation. Chapter X brings out case-studies of UN Operations where the air force blue marched shoulder-to-shoulder with the army OG. The results were for the world to admire. The last chapter delves into the gamut of training – a not – so glamorous role, yet more critical to a combat force's effectiveness than anything else. An appendix lists out the very-deserving presidential and gallantry awards of the helicopter fleet.

It is indeed difficult to get all primary sources of data, and therefore, the reliance on some secondary sources and personal accounts. At the end of each chapter or major operation, selected citations have been reproduced. These are all from the official govt records. They summarise and support the narratives far more effectively than any other source. It also brought goose-pimples when I was assorting them – the raw courage, professionalism and the will to do the soldierly act stands out abundantly.

This book is above all a tribute to the professionalism of a great service – the Indian Air Force.

CHAPTER I

Down Memory Lane

Birth and Growth

By the early 1950s, the Indian Air Force (IAF) had already embarked on an expansion and modernisation plan. The government had given its approval for induction of modern aircraft and equipment. Vampire jet fighters had entered the IAF in early 1948 and plans to induct the Toofani jets and C-118 Packets were finalised. It was at this time that helicopters emerged with the acquisition of a few Sikorsky S-55 and Bell-47 IIIG helicopters. 104 Helicopter Unit (HU) was formed on 10th Mar 1954 at New Delhi. The first S-55 Sikorsky arrived at Bombay by sea on 19th Mar, was erected in record time, test flown on 23rd, and ferried to Palam on 25th. The pioneers, Sqn Ldr SK Mazumdar and Flt Lt A Neal Todd, had just returned to India after training in the USA. These two soon started the process of training others – VK Sehgal, Jaswal, AS Williams, K Narayanan, B Johnson and RD Pant. Training was carried out on the lighter Bell helicopters, four of which were inducted in 1954 for training and recce & liaison duties with selected army formations.

104 HU was initially raised as a Helicopter Flight with Todd as its first CO, but was upgraded to a full-fledged unit on 1st May 1958 with Mazumdar leading the pack. During the very first year of its service i.e. 1954, the newly acquired helicopters earned great praise when they were called "Harbingers of Life" by the Indian press. Todd evacuated 15 villagers marooned in the Yamuna river near Delhi and started the long saga of disaster relief and mercy missions which IAF helicopter crews have been undertaking since then with determination and courage. On 23rd Nov 1959, 105 HU was raised and equipped with Bell-47 IIIG helicopters. Soon, both these unit were reequipped with the newly acquired Russian Mi-4s. On 1st Jan 1960, 107 HU was formed and equipped with S-55 and Bell-47 III Gs. On 26th Aug 1961, 109 HU was raised and equipped with Mi-4s.

Historically, helicopters had already arrived in the Indian skies during World War II in the Burma Campaign with a Sikorsky R-4 of US Army carrying out the first ever rescue mission by a helicopter. It was once again a Sikorsky (S-55) which first flew with IAF markings on 23rd March 1954. The Sikorsky S-55 was the world's first certified commercial helicopter. Within a few days of its arrival, it was tasked to carry the Prime Minister, Pandit Jawahar Lal Nehru to Tilpat to

witness a fire power demonstration by the IAF. The S-55 had the distinction of not only getting operationally deployed in NEFA but also being the first helicopter to be operated on board the aircraft carrier INS Vikrant. This small fleet was augmented by Bell-47s in 1957. These helicopters were smaller and were ideally suited for reconnaissance, training and rescue operations in mountainous terrain. The Russian Mi-4s were inducted into IAF on 23rd November 1959 at Jorhat. This also marked the beginning of a long line of Soviet era helicopters which continue to serve the IAF till date. It soon became a familiar sight in the far corners of India till phased out in September 1981 after 21 years of service. They provided a major capability of medium lift and remained the back bone of IAF operations till they were replaced by the Mi-8 helicopter in 1972. Meanwhile the lighter piston engine Bell-47s was soon complemented with the French Allouette IIIs in 1962. Post the 1971 conflict, the light utility fleet was once again given a fillip with the induction of LAMA helicopters, which were a derivative of the Allouette III. These were developed on the specific requirement of Indian Armed Forces. It went on to create helicopter history in high altitude flying and since then has the distinction of sustaining the highest battlefield in the world – the Siachen Glacier.

A helicopter training unit had been formed as far back as on 2nd Apr 1962 at Palam and was equipped with the Bell-47 III G helicopter. Later, this unit was renamed Logistics Support Training Unit on 1st Dec 1962, and was moved to Allahabad where Pilots Training Establishment was already carrying out flying training in basic stage on HT-2s. It was placed directly under HQ Training Command on 14th Nov 1966 and moved to Jodhpur on 22nd May 1967. LSTU then moved to Hakimpet on 15th Oct 1973, where it remains till today. It is the 'Alma-Mater' of the fleet and provides the much-needed firm foundation for this versatile fleet. The growth of the helicopter fleet also necessitated changes in the organizational structure at Air HQ. A Joint Directorate of Helicopters was formed under the control of Directorate of Transport and Maritime Operations. Today, there is a separate directorate for the transport and helicopter fleet headed by an Assistant Chief of Air Staff and the sub-directorate of helicopters has a dedicated Principal Director along with a host of directors and joint directors overseeing all aspects of the fleet.

Helicopters from Russia (or the former Soviet Union) continued to form the backbone of the fleet with the early eighties witnessing an upgrade drive. The Mi-25 was inducted in 1983 with the raising of 125 Helicopter Squadron. 1984 saw the induction of the Mi-26 into 126 Helicopter Flight operating the heaviest helicopter in the world with a lift capability of 20 tons. The very next year, the raising of 127 and 128 HUs marked the induction of Mi-17s. These were upgraded version of the earlier Mi-8s and had higher rated power plants and astounding amount of fire power. The early 1990s was marked with the arrival of Mi-35s in 104 Helicopter Squadron, and an effort towards self reliance took off with the Advanced Light Helicopter taking to the skies in IAF colours.

Operations

First Steps Two Bells of 104 HU operated in relief of marooned people from 28th Dec 1957 to 10th Jan 1958 in Ceylon (now Sri Lanka) when it was ravaged by floods. It was a 'first' that started the involvement on many occasions of IAF helicopters in humanitarian assistance and disaster relief abroad. In Nov 1961, M Boulet of SUD Aviation, a French test pilot, came to India for demonstration flights of Allouette-III helicopters. Sehgal of 104 HU created history by landing at 19,000 feet in the Himalayas. Mi-4 helicopters flew extensively in flood relief operations in the Brahmaputra Valley and later provided valuable support to the Indian Army against insurgents in NEFA. Flight Lieutenants Tandon, MS Sekhon, MN Singh, Sgt K Sharma, Cpl AK Sharma and NC (E) Papparao of 105 HU earned many honours for these operations. One of the original Mi-4 units (109 HU) had its helicopters operating in support of 17 Infantry Division during the swift police action in Goa during December 1961. It transported general staff officers, flew in vital despatches and evacuated the few casualties.

1962 The fleet had just arrived at the starting block when tension mounted on the border with China. The Indian government had decided to deploy army units all along the border to check intrusions; in an area that was remote, undeveloped and without any roads or railway. The task of maintaining the Indian Army was naturally given to the IAF. While fixed wing transport aircraft like the DC-3 Dakota, C-119 Packet and the An-12 quickly geared up to start air maintenance to forward airfields and Advance Landing Grounds (ALGs) either through para-dropping or air landed operations, the extremely important task of liaison, communications and casualty evacuation was the responsibility of the fledgling helicopter fleet. In Jul 1962, the Bells of 104 HU were actively involved in casualty evacuation in the Galwan valley of Ladakh. This was the first time that the Bells had operated above 10,000 feet AMSL. Throughout the period of tension and later the actual fighting in 1962, helicopters of the first three units tirelessly and fearlessly evacuated many a casualty and brought timely succour to the injured by way of urgently needed medical supplies, essential stores and equipment.

Countering the Chinese were inadequate troops of 4 Infantry Division; of which the forward brigade at Tawang was at the front line. Defensive build-up in the area had been limited because of the single, tortuous, road that wound its way for hundreds of miles through 15,000-ft mountain passes and forests up to the Namka Chu valley, and precariously past the Thagla Ridge. A handful of Bell 47s from 105 HU were attached to the 4 Div for reconnaissance and liaison flights, flying up and down the Namka Chu valley carrying Corps, Divisional and Brigade commanders. Two Bell 47s were located at the helipad at Zimithang, and were soon in use to lift casualties from the improvised helipads up at Lumpu and Tsangdhar. Four Mi-4s based at Tezpur, the HQ of 4 Corps in the plains of the Brahmaputra, took on Chinese ground fire while flying in reinforcements and supplies and evacuating casualties from the advanced landing pads. Hostile fire and poor weather conditions had to be tackled daily, but during the short breaks, Mi-4s were able to fly repeated sorties. Two Bell 47s at Zimithang and a number of Mi-4s were lost to enemy action in the Tawang area and in engagements further east in Walong.

On one such mission in Oct 1962, Sqn Ldr VK Sehgal became the first war casualty when his helicopter was fired at by the enemy in the border regions of NEFA. Sqn Ldr AS Williams, CO of 107 HU, was awarded the VrC for his heroic exploits. He had gone looking for Sehgal who had failed to return but was himself fired upon by the Chinese. With the oil-pipe line of his helicopter damaged, he cleared the ridge and flew safely back to a piece of small level ground beside a stream and was able to run back to his base. Soon the area was over-run by the Chinese.

With the formation of 110 HU on 10th Aug 1963, the IAF helicopter fleet had reached a sizable strength and was equally active in peace time duties.

1965 In early 1965, border incidents along the Cease Fire Line (CFL) in J&K were on the increase. Pakistan embarked on a probing action in the Rann of Kutch in early April. Two Mi-4s of 109 HU were soon operating from the ALG at Khawda for border recce and support operation during the Kutch incident. In July, a massive air drop of food and supplies was carried out for marooned troops in the same area. By August, six Mi-4s were sent to Jammu and later to Srinagar. This detachment was instrumental in helping the Indian Army check the influx of Pakistani raiders and infiltrators. This was the first time that these medium-lift utility helicopters were used in the offensive role. They carried out many sorties bombing and strafing the enemy raiders in the difficult and otherwise inaccessible mountainous terrain of J&K. This was in addition to their normal duties of supply dropping, casualty evacuation and chalking up on impressive tally of sorties in support of the Indian Army. This invaluable support finds an apt mention in 'War Dispatches', the story of the Indian Army's Western Command written by Lt Gen Harbuksh Singh.

During the misadventure of Operation Gibraltar by Pakistan in 1965, a helicopter task force, consisting of three squadrons was formed to assist in fighting against the Pakistani armed infiltrators who had entered Jammu and Kashmir in August. It was mainly based in Srinagar to carry out many offensive sorties against the infiltrators from 20th August 1965 till the end of hostilities. With suitable modifications, these helicopters bombed and strafed the positions of the raiders in many areas, especially Haji Pir Pass, Tangdhar, Badgam, Mandi, Budil, and the hills around Gurez. While these offensive sorties did inflict much damage on the enemy, more importantly, they exerted a great demoralizing effect on Pakistani guerrillas. They played an equally important logistical role by dropping approximately 92,000 kg of essential stores and urgently needed ammunition to army columns operating in different areas, lacking suitable ground communication. They also evacuated critical casualties from inaccessible areas; flying 198 trips, each loaded to maximum capacity. Some of these helicopters, including three Allouette, were used by senior army officers to get a good view of the areas of operations so that quick decisions could be taken to plan and execute counter-offensives against infiltrators.

1971 Mi-4 units saw continuous operational service in India's northeast, supporting the army's counter-insurgency operations in Nagaland and Mizoram, lifting commando teams into action and helping to pre-empt hostile infiltration to and from China, many times under hostile fire. The operational climax for the Mi-4s in IAF service came during the 1971 war when it played a key role in the Indian Army's Blitzkrieg-style advance to

Dacca in Bangladesh. Hostilities had broken out on 3rd December 1971, and a lightning Indian campaign was unleashed in the eastern sector.

While the Indian Army's 2 Corps battled its way from the western side, the 4 Corps thrust came from three directions in the east; across 250 km of the border between the erstwhile East Pakistan and Meghalaya to the north, through to the Feni Salient in the extreme south of Tripura. One column moved in from Silchar – Karimganj area towards Sylhet, another along the Akhura-Asjuganj axis and the third in the south with the objective of containing Comilla and cutting off Chittagong. Whilst battling against the two Pakistani divisions ranged against them, 4 Corps had to contend with the most formidable riverine terrain, criss-crossed by rivers and streams that make the area a logistic nightmare for any offensive action.

The advance was enabled by 110 HU, with 10 Mi-4s on strength and another two Mi-4s attached from 105 HU. The Mi-4s were intimately involved in the spectacular progress made particularly by 8 Mountain Division. It transported troops, guns, ammunition and equipment, evacuated casualties and conducted reconnaissance for the field formations in the sector. Between 7th and 10th December, more than 1000 troops and 300 tons of equipment were airlifted in the face of heavy ground fire for the the build-up against Sylhet, which was held by the Pakistani 313 Brigade. The helicopter task-force then moved to Agartala for the heliborne operations from Brahmanbaria to Raipura.

Mi-4s forced the pace of the Indian Army's relentless drive, which now saw heliborne-infantry leap frogging past natural obstacles and enemy defences, with MiG-21s and Gnats strafing and rocketing strongholds, in a text book display of co-ordinated air power. The biggest operational task yet undertaken by the IAF's task force was between 11th and 15th December. On the 11th, 1,350 troops and 192 tons of equipment were airlifted from Brahmanbaria to Narsingdi, including in particular the lead battalion 4th/5th Gorkha Rifles of 59 Mountain Brigade, spearheading 8 Mountain Division's intrusion into Sylhet. The Gorkhas and their equipment were lifted across the Meghna River to south-east of Sylhet, forcing the much larger enemy force to abandon defences and, for fear of being cut off by the heliborne forces, pull back into the town. On 14th December, the 12 Mi-4s were positioned at Dandkandi to airlift, across the Meghna to Baidya Bazar, 810 troops and 23 tons of equipment; and in another major effort the next day, they airlifted 1,209 troops and 38 tons of equipment to strengthen the Baidya Bazar complex. All told, some 4,500 troops and 515 tons of equipment were helilifted by the 12 helicopters in this action.

Well acknowledged by the Indian Army, IAF helicopters were a key factor in the lightning campaign, where the speed of advance was of the utmost importance. Other units that were given special mention were 15 and 24 Squadron whose Gnats provided escort to the heliborne operations and close support over the entire Brahmanbaria-Sylhet axis. Other Mi-4 and Alouette III units included for example, 112 HU in the Jessore sector, conducting casualty evacuation, communication and reconnaissance flights. Some Mi-4s and Allouette III were equipped as gunships and operated in this role both in the east and the north-west of India, recalling their earlier role in 1965, when they hunted infiltrators with machine guns and anti-personnel bombs in Kashmir.

Op Falcon/Chequerboard (1986-87) On 16th Jun 1986, the Chinese intruded into Sumdrung Chu valley in Arunachal Pradesh and started making permanent structures. This was discovered by 12 Assam Regiment when they returned after the winter withdrawl. Some 200 Chinese soldiers continued to illegally occupy despite formal protests by the Govt of India. By August, they had constructed a helipad and air supply had started. In response, an entire brigade of 5 Mountain Div was airlifted by IAF Mi-8s and the newer more powerful Mi-17s to Zimithang which overlooked Sumdrung Chu. In response to massive Chinese mobilisation from Chengdu and Lanzhou, the Indian Army and IAF launched Op Chequerboard to move 10 Divisions and a number of IAF fighter and helicopter squadrons. Three divisions of Indian Army mobilised in the Arunachal sector were entirely supplied by IAF Mi-17s till mid-1987. This was a herculean effort from the IAF's helicopter fleet which is not very well documented or acknowledged.

Op Pawan Another major operation which saw extensive employment of IAF helicopters was in Sri Lanka. From October 1987 to March 1990 when the IPKF finally returned to India, the helicopter fleet was used in large variety of roles. While the most important and extensive was airlift of supplies, troop lift and reinforcements, the Mi-8s were also used in the offensive role. These specially armed helicopters delivered the deadly cargo of rocket, bombs and other armament accurately against the numerous jungle hideouts of the LTTE. As in 1971 Bangladesh, here again IAF helicopters gave invaluable support to the Indian Army (IPKF) in crossing the numerous water obstacles and picked up troops or air landed them in remote areas of the thickly wooded terrain of northern Sri Lanka. With their support and maintenance base at Sulur, the Mi-8s, Mi-17s, Chetaks and also the Mi-25 gunships maintained detachments at Jaffna, Trincomalee, Batticaloa and Vavunia. Hundreds of missions were executed in the face of sustained and deadly LTTE fire in many daring joint operations.

During the nearly three years of continuous operations, IAF helicopters flew thousands of hours and lifted numerous troops and evacuated a very large number of casualties thus boosting the morale of the fighting soldier. A large number of aircrew, both pilots and engineers, earned Vir Chakras, Vayu Sena Medals, Yudh Seva Medals and mentioned in dispatches. One Mi-8 unit alone flew over 3,800 hours and airlifted 42,000 troops during Op Pawan. In the post IPKF period, attack helicopter units trained hard to man the newly acquired Mi-35s, an improved version of the Russian Mi-25 gunship. In retrospect, it can be stated with justifiable pride that the helicopter offensive arm of the IAF had come of age and had proved itself as an indispensable element of the awesome air power arsenal of the nation.

1999   During Op Safed Sagar, for the first time Mi-35s were positioned in the Srinagar valley across the Banihal pass in Jul 1999. Pakistani rangers had occupied the heights in Mushkoh valley; Dras and Kargil and were in the firing range for Srinagar-Leh highway. The Indian Army was given a green signal to go ahead with military action on 20th May. On 26th, Mi-17 helicopters and Mig-21s struck the heights of Tololing and Tiger Hill overlooking Dras. The Mi-17 fleet played a key role in logistics support and under slung operations of artillery guns across treacherous terrain and valleys. During this period, 104 HU was tasked to sanitise Srinagar and Awantipur airfields and also employed against ground threats in Srinagar valley. For the first time the attack helicopter units carried out practice firing at Toshe Maidan firing range, south-east of Srinagar airfield at an altitude of 10,000 ft. Subsequent modifications and trials allows the Mi-35 to conduct operations at higher altitudes.

UN Peace Keeping

Somalia   The IAF took part in peace keeping duties in Somalia from 1st Oct 1993 to 21st Dec 1994 as part of the Indian Contingent supporting UN operations. On 22nd Apr 1992, in response to a recommendation by Secretary General, the Security Council adopted resolution No.751 (1992) by which it established UNOSOM. The first batch of troops landed in the capital city of Mogadishu on 28th Aug. The induction was completed by 22nd Oct. The Indian hospital and aviation unit was operational immediately. IAF helicopters in Somalia as part of UNOSOM were utilised for the following tasks: road opening and convoy escort; aerial reconnaissance; casualty evacuation; communication.

The first operational mission was undertaken on 12th Oct 1993 with the Bde Cdr and his deputy on board for a recce in and around Baidoa. In Feb 1994, the contingent gave aerial cover to the Mahar Battalion who were escorting 500 refugees to their villages. On 8th Dec, a RPG fired by Somali militia during inter-clan fighting exploded on the roof of the barrack housing IAF officers. In the explosion two officers and an airman were injured. While western nations used force to achieve a secure environment and failed, the Indian Contingent used a more humanitarian approach. They took pains to learn the Somali culture, customs, and language and respected their religion. Without taking sides, all conflicts in the Indian area of responsibility were settled amicably by discussion rather than force. The IAF took part in a UN peacekeeping mission after a gap of 30 years.

Sudan   The Indian Aviation Contingent was the first military aviation contingent that was deployed in United Nations Mission in Sudan (UNMIS) in October 2005. The contingent came to Sudan as part of UNMIS subsequent to passing of UN Security Council resolution 1590 on 24th Mar 2005 under Chapter VII i.e. Peace Enforcement. The large scale humanitarian crisis in Sudan due to the armed conflict led by leaders of north and south Sudan over political control over rich natural resources had led to large scale refugee problems and associated malnutrition and hunger deaths; the resulting situation necessitating an intervention by the United Nations. Consequent to the UN resolution, six MI-17 helicopters of the IAF were deployed at Kadugli in central Sudan. Five contingents completed their deputation of one year each. The last contingent was repatriated on 27th Dec 2010.

IAF helicopters in Sudan as part of UNMIS, and in support of the Indian Army and other ground troops, were utilised for tasks such as: Insertion/ Extrication of Quick Reaction Force (QRF); Underslung Operations; Air Patrol/Observation Missions; Carriage of Cargo; Medevac/Casevac. They chalked up 10,420 hours of flying, 14,35,814 kg of load lifted and 71,814 passengers transported during the deployment. The professionalism and involvement shown by the personnel of the Indian Aviation Contingent in discharging their duties to bring peace and harmony in Sudan has not only done the IAF proud but also brought laurels to the nation.

Democratic Republic of Congo (DRC)   Indian Air Force Contingents (IAC) in the Congo truly stood tall among all others in the business of peace. Those who were there in DRC from across the world from 2003 to 2011 have acknowledged publically that the will and professionalism displayed by the IAF was second to none. Even the Indian Army, deployed since 2005, has gone through its ups and downs, including scathing criticism since 2007. But report after report, UN-sponsored or independent, concedes that none ever found IACs wanting in robustness. This speaks very highly of the Indian Air Force in terms of ethos, ethics and air warrior-like qualities. While all IACs have repatriated from UNPK, bringing to an end a fruitful chapter, this is an unsung but very important experience in peacekeeping history of the UN.

MONUC was the third major contribution of the IAF under UN flag, after Somalia and Sierra Leone. As the situation in North-East Congo turned grave with repeated massacres and killing of innocent civilians, the international community decided to strengthen the military presence. India contributed armed helicopters and utility helicopters in the Congolese provinces of North Kivu and Ituri. The IAF unit in Goma/Bunia was called the Indian Aviation Contingent (IAC-I) and its major lodger units included IAF Squadron 2003 (Vipers: Mi-25s) and IAF Squadron 2004 (Equatorial Eagles: Mi-17s). Vipers were equipped with four Mi-25 aircraft while Equatorial Eagles had five Mi-17 helicopters along with aircrew and ground support personnel. These invaluable assets increased MONUC's sphere of influence in the Eastern DRC; the UN forces were able to reach areas that had so far been outside its sphere of influence.

Indian Aviation Contingent-II (IAC-II) was inducted in Feb 2006, with its fleet of six night-capable Mi-17 helicopters at Kavumu Airport in Bukavu and four night-upgraded Mi-35 helicopters at Goma, alongside IAC-I. Though it was not the first time that Indian Air Force was fielding a contingent abroad, it was certainly the biggest, with 10 helicopters, 36 vehicles and initial strength of 285 personnel. The first batch of the IAC-II Contingent (Rotation-I) arrived at Bukavu on 26th Feb and became operational in March. As the scope of operations increased in the eastern DR Congo, UN's requirement of air effort increased at Kalemie base in Tanganyika region, and a new detachment was started.

The Future

Low intensity conflicts and short duration highly confined but fast and intense conflicts are more likely than wider conventional wars. Though airpower would play a vital role, the boots on the ground and combat enabling missions of airpower would form the backbone of any success. The helicopter was perhaps tailor-made for mountain operations and thus will play a pivotal role, especially in providing tactical mobility, logistic support and fire support besides numerous other support roles. Keeping in mind the resource constraints infrastructure and cumulative skill level, the IAF is best placed to operate the large inventory of light, medium and heavy types that is required in the future.

At the same time, a strong case exists for greater jointness in operations. Scenarios of LICO encompassing counter-insurgency, counter-terrorism or even localised wars like Kargil might evolve in the future in J&K, Nepal,

Bhutan, Myanmar and NE India. Some of these could be fanned by adversaries and competitors like Pakistan and China. Highly mobile light infantry and Special Forces adequately supported by fixed wing and helicopters may help contain them from growing into more serious confrontation. In fact, given the scale of involvement and risks that would be associated with helicopter operations in future conflicts in the Himalayas, combat support is a misnomer-it is very much part of combat operations. Lessons from recent campaigns clearly suggest a revamp of man, machine and tactics.

IAF Helicopters at this point are well poised to take on the emerging threats of modern warfare. Their experience during Op-Pawan in mid-eighties, and thereafter the involvement in Kargil and anti-Naxal operations has set think-tanks in the IAF to debate the role of helicopters in high intensity and intensely fluid future conflicts. Helicopters of the IAF would play a dominant role in facing challenges of Fourth Generation Warfare. While the debate continues towards defining the role of helicopters in such non-traditional threat environment, the need to continuously upgrade the equipment and capability was never lost sight of. A strong case existed for quantitative and qualitative increase in light, light combat, medium and heavy lift helicopters in the IAF inventory.

Helicopters inherently are capable of multitasking and provide critical assistance in border infrastructure developments, disaster management and in a number of tasks in the form of 'Aid to Civil Power'. In terms of pilot training, the need to look for intense but low cost options like simulators, enhanced night capability and special skills which improves the ability of aircrew to adapt to new and changing situations is a work-in-progress. The helicopter fleet is already under a major transformation with induction of Mi-17 V5. It is envisaged that the contract for 22 attack helicopters, in which the AH-64 is the front runner, would soon be awarded. In addition to these, contracts for heavy lift helicopters and light utility helicopters are also likely to be declared shortly. The VVIP Communications Squadron is also poised for a major upgrade with the induction of Agusta Westland AW-101. Indigenisation will get a boost once the HAL-made Light Combat Helicopter is inducted into the IAF. Training of the fleet has kept pace with induction of full-motion simulator programmes and centres of excellence for tactical training.

Induction of the Modern Mi-17 V5 on 17 Feb 2012:
Raksha Mantri, Chief of the Air Staff & Vice-Chief

CHAPTER II

Lessons from Recent Campaigns and Operations

With the world in recession and defence budgets under pressure, air forces are increasing their investment in multi-role capable helicopters. The combat landscape will encompass increased counter-terrorism operations that will require deployment of resources for law enforcement and internal security roles, as well as conventional battle capabilities. Asymmetric tactics in current conflict zones, such as roadside bombs used by insurgents, have made helicopters a 'must-have' for troop movement. These requirements are best served by helicopters with the ability to perform a variety of functions, move swiftly between locations, and switch roles quickly. Future conflict is likely to comprise of 'hybrid' threats that will require flexibility of response from a coordinated helicopter command with sufficient resources. Helicopters are the only resource with the flexibility to swing between every function required in the modern battlefield: ISTAR, attack, sustainment, troop-carrying, reconstruction and humanitarian action enablers. Their adaptability can also meet these needs across a range of terrains, moving quickly in and out of mountains, jungles, deserts, urban or rural environments. These qualities of adaptability and flexibility are critical in the 'hybrid' campaigns of the future.

The demands of a helicopter-force capability include: powerful vertical performance to maximise the flexibility and reach of ground forces; hot and high performance; operations in all environmental conditions, including snow, ice and dust; low-level flying and tactics; selective armour, self-sealing fuel tanks and redundancy systems; and finally, aircraft maintenance and operational sustainability must be possible in the harshest of conditions. Helicopters must allow rapid deployment of firepower and ground manoeuvre forces almost anywhere and even in the most difficult terrain. They would allow an unprecedented speed of insertion, resupply of weapons and other resources, as well as a swift withdrawal from the battle area if necessary. An important capability is to boost the morale and confidence of ground forces by rapid medical evacuation from the front line. They need to possess capabilities to access remotely deployed or isolated ground troops; and, overcome blocked or cut-off main supply routes or lines of communication. They must be able to lift troops, conduct casevac, datalink with UAVs and ground forces, pass real time full motion video, and bring suppressive fire-support if necessary. In the future, these aircraft will need to be able to swing between roles and operate across different environments more readily, including switching between sea and land.

A World Scan

When air power came into being in the 20th Century, it was armies that controlled this new dimensional capability shift. As the fledgling component grew in leaps and bounds due to technology jumps, armed forces evolved to have specialist air forces. It was driven by the need to understand and employ air power correctly, which could only be done by those whose profession, careers and lives were mainly devoted to the third dimension.

In the 1972 war over Sinai, the Egyptian Army lost more than 14 Mi-8 helicopters of the army while attempting a bold and audacious Special Heliborne Operations involving a train of helicopters. The entire armada was short-circuited by just two Phantom fighter jets of the Israeli Air Force. This was another case of ground commanders acting in isolation without specialist air force personnel on board the planning process. In comparison, the IAF's campaign of multiple SHBOs with Mi-4s in 1971 in (then) East Pakistan was a resounding success. The environment was suitably shaped by the IAF with reasonable air superiority and strafing of ground threats by Gnats and Hunters. Similarly, a very successful employment of Apache attack helicopters in the 1991 Gulf War in taking out various targets such as ground radars, armoured formations, missile sites etc. was a result of good and effective integration with the USAF's overall plan. In their effort to carve out their own legacy independent of the USAF, the US Army made a blunder on 23rd Mar 2003, when 30 Apaches went out of action in a bold attempt to take on the Iraqi Republican Guards. The same error was repeated in Op-Anaconda in Afghanistan during Enduring Freedom when seven Apaches were put out of action. The disastrous situation was somewhat salvaged by the USAF finally joining in the fray three days later.

These examples, and more that follow, serve to bring out an important lesson being learnt the world over – that effective integration of joint capabilities is the way-ahead. While turf-wars and differing viewpoints will always be there in good measure in any country, evolved armed forces put their money on the jointmanship horse rather than going it alone. A scan of the armed forces of China, Pakistan, UK, USA, Russia, Singapore, Israel and other brings out different models of effective integration that have been adopted. The Chinese Air Force or PLAAF was for a long time relegated to the background. The first Gulf War and advances in technology and concepts jolted the Chinese to the reality facing them, especially across the Taiwan Straits. Subsequent investments into the air force and navy show a completely different thought process. Unlike India, the entire gamut of airborne forces including troops comes under the PLAAF. The nascent attack and medium-lift helicopter force has been put under the PLA.

The US has the resources to virtually fund entire independent helicopter assets and forces for each of the services– Army, Air Force, Navy, Marines and Special Forces. However, in the campaigns in Iraq and Afghanistan they had to relearn basic lessons on the costs to be paid in case any service tries to go it alone. The Russians originally had helicopter assets (attack and medium utility) under the army command. However, lessons during the Soviet deployment in Afghanistan and graver lessons in the First Chechen War convinced them that the air force understood and handled air power the best especially in a complex integrated scenario. For the second Chechen War in 1991, the control of helicopters moved to the air force commander (VVS) in the joint set up. Currently, they are evolving into true integrated regional commands, yet the basic premise has not changed. Israel and Singapore have the helicopter assets as a part of the air force while robust joint structures integrate the whole package quite effectively. The larger lesson is that all countries have unique models of capability building and even more unique ways of synergising and integrating. The bottom line is optimization, effectiveness and avoiding redundant duplication. More so today than ever, the stress is on cost-benefits analysis and avoiding huge initial and recurrent costs.

The Indian model, enunciated by the visionary Joint Army Air Force Integration Implementation of 1986, has truly stood the test of time. The IAF's helicopter fleet has not only been a great partner to the Indian Army's endeavours in counter-insurgency action at home in J&K and Northeast, but has also been a great enabler and force-multiplier abroad in UN missions and in the IPKF saga. At the same time, as is being realised by NATO and others, the fleet serves the nation during peace time in multifarious roles such as infrastructure building, aid to civil authorities, action with severe restrictions on collateral damage, humanitarian assistance and disaster relief. The IAF capability has demonstrated the flexibility to switch roles with great adaptability. The examples of campaigns that follow underline the flexibility and adaptability that is required by modern armed forces.

UK's Joint Helicopter Command (JHC)

The JHC comprises battlefield helicopters of all three Services of United Kingdom and a potent air assault capability in the shape of the 16th Air Assault Brigade. The JHC has been on continuous operations since its inception in 1999 having emerged as a consequence of the UK's Strategic Defence Review. Having learnt its lessons in Bosnia and Sierra Leone, the JHC has really come of age since 2003 in operations in Iraq and Afghanistan. In Basrah, the UK combined the efforts of RAF Puma, Merlin Mk3s and Chinooks, RN Sea Kings, and RN and Army Aviation Corps Lynx from the same airfield, with common operating procedures, a single point of command (rotated through each Service) and all sharing the same accommodation, briefing, engineering, eating and relaxing facilities. The RAF's doctrinal foundation of 'Centralised Command for Decentralised effect' holds absolutely true for the joint battlefield helicopter operational and non-operational output of the JHC, with a joint force in theatre, and a co-ordinated and unified joint command at home. The intense operational tempo of Iraq and Afghanistan, combined with the requirement to deliver such a broad range of aviation, could not be achieved without the Joint Helicopter Force.

According to UK's MoD, in the future, joint helicopter forces should be fully capable of providing real-time intelligence collection, surveillance and reconnaissance to ground forces whilst also being confident in their self-protection from small-arms, rocket propelled grenades or manpad attacks. However, they should be capable of swinging from the intelligence, surveillance, target acquisition and recce, or ISTAR role, into an interdiction or attack role, a sustainment and troop-carrying role or perform aid to civil authorities or humanitarian operations. As the sophistication and complexity of the threat in both Iraq and Afghanistan have increased, so has the tasking of aircraft and demands on aircrew to keep pace. As the skills of aircrew have developed, their training and core-skills have been raised to meet the complexities. What is most significant is that battlefield aviation, and specifically the dynamism created by helicopters, have become an absolutely essential capability for the land commander across the full spectrum of operations, from humanitarian relief and stability operations, through counter narcotics and counter-insurgency operations, to full scale war fighting.

US Army Operations

U.S Army Aviation: Bosnia   In the 1999 Kosovo operation, the Army Aviation Task Force (TF-Hawk) was sharply criticized for its training, personnel and equipment readiness, deployment time, and the size of the overall package and number of C-17 platforms required for its airlift. What deployed to Albania in support of Operation Allied Force was in reality an oversized army task force, which was larger than the support force required for using Apache helicopters in military operations. If the Apaches had been fully integrated into JFACC operations and based with other forward deployed assets, the force could have been much smaller. The air movement of a self-deploying army aviation task force of 61 helicopters and the logistics tail required for their command and control, maintenance, security and support required *442 C-17 sorties* for deployment and sustainment of Task Force Hawk. And, as was inevitable, due untenable high risk in the existing environment, the task force never saw action – a complete waste of effort because of myopic vision.

U.S Army lessons: Iraq During the approach to Baghdad on 23rd Mar 2003, ambitious deep-shaping operations by the US Army went awry when some 30 Apaches got hit and went out of action. One set of crew was captured while the entire attack helicopter (AH) battalion was out of action for the rest of the war. The Iraqi Republican Guards had set up an ambush activated by a cell phone network, a classic asymmetric response to classic conventional overreach. Thereafter, AH action continued all throughout the war except that tactics were changed and a combined-arms approach dovetailed into the overall USAF plan was adopted. Fixed-wing and artillery would take out the ambush sites before the Apaches moved in for the kill. The same lesson had been learnt by the Soviets in Afghanistan and the Russians after the First Chechen War.

U.S Special Operations Lessons learned by US special operations forces (SOF) over the last two decades and demonstrated in Afghanistan provide some signposts for future conventional forces and the need for effective integration. USSOCOM and its predecessors have spent the last 20 years forging a joint team with interoperable service components. SOF personnel jointly conduct virtually all training above the individual skill level. This training program is tough, extensive and expensive, but it has succeeded in forging a truly interoperable team. SOF communication link SOF service components-and extend to parent service forces as a result. SOF personnel conduct operations with elements from all services directly integrated in tactical formations – from SOF or SEAL teams with integral Air Force air-control elements to tactical helicopter formations combining Army and Air Force aircraft. SOF personnel have proven uniquely suited for this networked, distributed warfare. Special Forces (SF) teams with embedded Air Force air-control elements provide a tactical force with a broad range of skills and the maturity to execute mission orders without detailed oversight. They can move, shoot, and communicate while employing supporting fires from any source-land, sea, or airpower from US or coalition forces; essentially because of interoperability. They truly epitomise effective integration of capabilities.

U.S Army Helicopters in Afghanistan   U.S. helicopters suffered from high-altitude capability limitations similar to those faced by Indian aircraft at Kargil. Just as Indian helicopters provided invaluable transport capability to the Indian Army, the CH-47D Chinook proved to be a crucial source of mobility and heavy lift for U.S. forces in the mountains of Afghanistan. Unlike Indian forces at Kargil, the U.S. military was able to employ its heavy attack helicopter, the AH-64A Apache. Altitudes approaching 18,000 feet at Kargil had prevented the IAF from using the Mi-25 attack helicopter, forcing reliance on the lighter Mi-17. Seven AH-64A Apache attack helicopters escorted the CH-47D Chinooks that ferried soldiers and supplies to the landing zones (LZ), and remained on station to provide cover for the infantry.

The superior Apaches could not hover in the rarefied air, and had to rely on 'running gunfire' to engage targets, similar to the Mi-17s employed by the IAF. But mountainous terrain made the superior Apache equally vulnerable to enemy fire. Al Qaeda positions were well hidden in irregular terrain, similar to the NLI positions at Kargil, forcing the helicopters to fly as close as 200 meters to enemy positions to identify and engage them with fire. Most fire against U.S. helicopters consisted of rocket-propelled grenades (RPG) and machine guns, for which there is no countermeasure. Mountainous terrain further limited the Apaches' effectiveness, disrupting line of sight radio communications and making coordination with other aircraft and ground forces difficult. Five of the seven Apaches were eventually disabled by enemy ground fire and forced to withdraw. Four U.S. Marine AH–1W Super Cobra attack helicopters augmented the force, and performed well, ultimately completing sorties in close support of ground forces; thereby making the case for simpler and uncomplicated machines even stronger. Also, the disastrous situation was only somewhat retrieved by the heavy suppressive support of the USAF, though asked for at a very late stage.

Soviet Helicopter Tactics

The Afghan Challenge: A Constant Battle for Evolving Better Tactics   The genesis of the Afghan conflict lies in the late 1970s when the pro-Soviet regime in Kabul came under severe threat from the Mujahideen, a mix of Afghani, Iranian and Pakistani militia with tacit backing from the U.S.

Perhaps more than any other modern conflict, the world saw the Afghan war as a helicopter war. The Kabul regime, used to operating Soviet helicopters since the 1950s, was provided with the export versions of the Mi-25 and the Mi-35. Pitted against the Russians were the Afghani rebels with predominantly land forces and no air assets worth the name. The threat posed to the Soviet helicopters were: light, medium and heavy machine guns; RPGs; anti-aircraft guns; SA-7 SAMs (basically 1983 onwards); and the Blowpipe and Stinger-1986 onwards.

Being no match in a conventional engagement, the Mujahideens adopted "hit and run" guerrilla tactics. As a result of the peculiar Afghan terrain and the tactics adopted by the rebels, the Soviet army's mechanized combined armed forces found they could not move into action quick enough to intercept the Afghan resistance. The usage of artillery as a weapon of choice was found wanting as they required to be set up, and thereafter secure firebases. Helicopters were used to provide mobile fire power in CAS operations, attacks on rebel supply convoys, positions and staging areas as well as for the whole gamut of independent operations, in support of combined arms ground ops and special operations forces ops. For these missions, the Hind was armed with 12.7 mm Chin gun or 30 mm cannon, 57mm or 80mm rocket pods, high explosive (HE), white phosphorous or incendiary bombs, " liquid – fire" delayed action incendiary canisters and Swatter, Sagger and Spiral antitank missiles.

The first Soviet use of armed helicopters in significant numbers and the first Soviet helicopter tactics both dated from 1967, only 12 years before they committed in Afghanistan. Therefore, there was little background available for the operations undertaken and tactics to be adopted. The SOP was to engage targets at low speeds and attitudes carrying out dive attacks with machine guns, 57 mm rockets, cluster and 250 kg HE bombs. They would commence their dives at greater than 3000 feet and end with a sharp evasive turn or low level repositioning for a repeat attack. This was because the threat was predominantly small arms.

But with the Afghan resistance getting stronger and helicopter losses mounting, the Russians undertook a change in tactics visible early-1980 onwards. Self preservation and the realisation that the war was going to be long-drawn motivated many of the changes. By 1981 itself, the Russians started using NOE flight patterns, flying 5-10 metres above ground. Simultaneously, crew protection was enhanced by way of armour plating against small arms. Since the start of the war till 1986, the Russians also followed the tactic of 'Fishing with Live Bait' i.e. sending one helicopter in at high altitude to draw fire while his wingman remained low, behind a ridgeline ready to attack whoever opened fire. These were pre-SAM days.

The use of helicopter 'Scouts' commenced with the Hips and Hinds with an experienced pilot on board who guided the attacking helicopters to the target while maintaining a high perch away from the action. The Hinds also undertook convoy escort role providing a helicopter escort of normally two to four Hinds overhead or in front of a column of vehicles. These would strike at suspected ambush sites, six km ahead of the column, firing an hour ahead of the column's arrival. The Rebels would however simply wait well back till the convoy was almost level with the chosen ambush site before quickly moving into position. Convoys struck thus would call for the Hind escort for direct support, which the gunships provided by firing against the roadsides.

The early 1980s saw the Russian helicopter undertaking a variety of roles ranging from scout to convoy escort and resorting to NOE flying and better armour as protection against the small arms & AA guns. However, the ground rules changed with the induction, by Mujahidins, of the SA-7 SAMs. This led to the Hinds and Hips ejecting flares from automatic dispensing systems during high-threat segments of sorties and making spiralling approaches, descending within the airfield's perimeter. In addition 'LIPA' IR jammers were employed and shrouding systems were fitted to exhausts to reduce IR signature, a further protection against the SA-7. In response to the new and improved Afghan AD weapons, new Hind attack patterns were adopted involving running in at low altitude starting some four to five miles from target and climbing to firing height just short of firing point to fire followed by a quick ultra low get away. A more cautious approach involved not closing within a mile of the target if it was known to be heavily defended with heavy machine guns or more.

Introduction of the Stinger   The nature of the conflict for the helicopters however, changed dramatically in Sep 1986 when three out of a flight of four Soviet Mi-24 Hinds were destroyed in quick succession by Stinger man portable, heat-seeking SAMs. The arrival of the Stinger led to a combined Soviet response on three fronts – technical, tactical and operational. On the technical front, the Russians responded with passive countermeasures using systems of baffles, well faired, and fitted around the exhausts of both Mi-24 and Mi-8/Mi-17. Cover plates placed at air intakes helped reduce IR signature in the forward hemisphere. Active measures adopted included IR decoy flares normally dispensed at two second intervals when taking off/ landing or flying in a high threat area. However, the flares could not provide all angle coverage as they fell down and below the aircraft.

'LIPA' IR jammer was fitted onto Soviet helicopter spines. These emitted IR frequency to barrage the missile's seeker head to prevent a lock. However, the Afghan pilots were quoted as saying that while this worked against the SA-7 or the U.S. Redeye, its efficacy vis-à-vis the Stinger was suspect. By 1987, the Soviet helicopters and Afghani Hinds were fitted with 'LIP' (a pulse Doppler missile warning system mounted under the tail boom). The cockpit display would show the bearing and range of an incoming missile, so evasive action could be taken. But its key drawback was its orientation to detect missiles only from below and to the rear while the Stinger could come from any direction. The Afghan topography involved flying down valleys, below crest line and the SAMS could come from crests both above as well as below. This was a change from the SA-7 threats where the missile could acquire the target only against the blue sky background. The technical response to the Stinger thus proved inadequate thus leading the Soviets to shift emphasis onto tactical counter measures.

The most obvious tactic against the Stinger was to fly above or below the missile envelope. The high altitude solution was however not feasible for the heavily laden helicopters. Thus the helicopters followed extreme low level flight patterns often along roads at 5 metres above ground or less. This 'road running' became widespread by 1987 because most roads in Afghanistan follow valleys and the Soviets usually established free fire zones along them, making the helicopter difficult to ambush. Still, the predictable flight path made the helicopter vulnerable to RPG-7 anti-tank rocket launcher attacks by Afghan guerrillas that lay in wait in roadside ditches. During all this, night operations gained importance and represented the most important tactical adaptation to the Stinger threat. The helicopters were flown on low-altitude contour flight patterns for insertions and extractions.

One of the most striking examples was during Op-Magistral – the relief of the besieged garrison of Khost in Dec 1987, when the Soviets successfully inserted a battalion sized air assault force using Mi-17s escorted by Mi-24s. This involved flying through mountains at night without lights and risking losses. Such sorties subsequently became a rule rather than an exception. The insertions were accompanied by suppressive arty fire including long-range fire-strikes from 220 mm BM-22 multiple rocket launchers, along much of the penetration route. Additional measures against the Stinger involved improved intelligence gathering and air recce resulting in the Soviets compiling computerized databanks of suspected & likely stinger concentrations. Cash rewards of 1,50,000 Afghanis were offered to pilots reporting a site. Maximum up-to-date information was received from human sources.

Larger Reasons for Soviet Losses   Soviet commanders, staff, and individual crews made serious errors that led to unnecessary casualties.

During the time in Afghanistan, they lost 329 helicopters, which included 127 helicopter gunships, 174 armed helicopter transports, and 28 lift ships. These significant losses were due to poor reconnaissance of the enemy and his air defence systems, poor command and staff work at all levels in organising and conducting combat, insufficient preparation of replacement flight personnel arriving from the Soviet Union, and the excessive overuse of army aviation. The demand for army aviation helicopter support grew significantly from year to year. This resulted in excess growth of the sortie rate for flight personnel. The average sortie rate for pilots reached six to eight flights in a 24-hour period and 600 to 800 flights in a year, with over 1000 hours of combat flying time. In this period, the flying crews displayed various signs of fatigue, disorientation, and the breakdown of the cardiovascular and motor systems, and frequently displayed pronounced psychiatric breakdowns. When a helicopter crashed or was shot down, the chances of trauma and death of the flying crew were often higher due to the ineffective use of rescue equipment; the poor survivability prospects if the crew cabin, central section of the fuselage, hydraulic system, or fuel system were damaged; and the poorly designed crew seats and seat belts, which, during emergency landing, would break.

Much has been written in the Western press about how the introduction of the U.S. Stinger shoulder-fired air defence missile won the war for the

Mujahedeen. The Stinger was an effective system, but an examination of Soviet aircraft losses shows no appreciable rise in the number of aircraft shot down after the introduction of the Stinger. Stinger did not shoot down that many aircraft. What the Stinger did was to cause a complete revision of Soviet aerial tactics. Once it was in theatre, helicopters stayed over friendly forces and limited daytime flights, jet aircraft flew much higher, and all aircraft took electronic and other countermeasures to survive. Stinger was effective – not by the number of aircraft that it downed, but by the change in tactics it engendered. Stinger made pilots cautious and less of a threat to the Mujahedeen.

Soviet Assault Tactics in Afghanistan   Air assault tactics and helicopter gunship tactics changed and improved steadily throughout the war. However, the Soviets never brought in enough helicopters and air assault forces to perform all the necessary missions and often squandered these resources on unnecessary missions. Helicopter support should have been part of every convoy escort, but this was not always the case. Dominant terrain along convoy routes should have been routinely seized and held by air assault forces, yet this seldom occurred. Soviet airborne and air assault forces were often the most successful Soviet forces in closing with the resistance, yet airborne and air assault forces were usually under strength. Air assault forces were often quite effective when used in support of a mechanized ground attack. Heliborne detachments would land deep in the rear and flanks of mujahedeen strongholds to isolate them, destroy bases, cut LOCs and block routes of withdrawal. The ground force would advance to link up with the heliborne forces. Usually, the heliborne force would not go deeper than supporting artillery range or would take its own artillery with it. However, the Soviets sometimes inserted heliborne troops beyond the range of supporting artillery and harvested the consequences. And, although the combination of heliborne and mechanized forces worked well at the battalion and brigade level, the Soviet preference for large scale operations often got in the way of tactical efficiency. Ten large conventional offensive involving heliborne and mechanized forces swept the Panjshir Valley with no lasting result.

Russian Experience in the Caucasus   The problems of helicopter support in the mountains of Southern Chechnya provide stark examples of the failure of command, pilot error and equipment failure throughout both the Chechen conflicts over more than a decade. Much of the poor state of military helicopters was due to a lack of investment in and direction of the military industrial complex dating from the Yeltsin years. The Russian defence set-up had neglected a viable long-term procurement option even when short-term procurement had been ruled out.

There were 18 helicopter losses in Chechnya and Ingushetia between August 2002 and April 2007. Ten crashes were either due to technical faults or pilot error, and only eight attributed to enemy action such as strikes by missiles, ground fire or grenade launcher. One of the worst helicopter disasters suffered was the downing of a Mi-26 at Khankala on 19th August 2002 with a loss of 127 lives. One of the factors which contributed to this disaster was the singular failure of command, the lack of discipline and control throughout the chain of Army command. Retribution followed swiftly, and General Pavlov, commander of Army Aviation was dismissed from his post and the ASV transferred back to the VVS (Air force). Russian helicopter experience in the Caucasus established that there was a need for gunships like the Ka-50 or the Mi-28, which can locate and engage targets from a safe distance, at night and in any weather. There were five significant operational lessons learned from helicopter fire-support operations in Chechnya according to Russian analysts:

- Enhanced target acquisition and PGMs are required to reduce collateral damage.
- Pilot proficiency is central in alleviating risk and improving capability. Also, helicopter gunships are becoming too vulnerable to operate in some environments independently. Rather, helicopter gunships should support land force elements in an integrated plan.
- The intelligence provided by UAVs can be effectively utilised by helicopters.
- Night operations functionally dislocated the Chechens insurgents. Hence, night vision equipment is a force enhancer.
- Close Air Support must be prompt, otherwise targets can escape.

Comparison of Chechen Conflicts (1994-1996 & 1999) The establishment of a unified air component of all air assets in the Second Chechen conflict improved coordination and cooperation and thus the effectiveness of airpower. Also, air support for ground forces operations was more successful in the second conflict. By conducting air barrages prior to the advance of troops, airpower created favourable conditions for ground forces and diminished the possibility of friendly fire. FACs proved to be more effective than in the first conflict. Because of their greater numbers, they could be deployed in more units and at lower tactical levels, sometimes even at company level; and were apparently better trained and better equipped with more sophisticated communications instruments.

Another ground for improved air support for ground forces operations was the formation of Aviation Tactical Groups. Combining target-designation and attack helicopters in tactical formations proved to be highly effective. In the First Chechen conflict, helicopters were mainly used for supporting tasks and were excluded from urban areas for fear of enemy air defence. It was then thought that for combat tasks, fixed wing aircraft such as the Frogfoot, would replace rotary wing. However, in the Second Chechen War, helicopters were fully involved for combat missions, which broadened the scope of airpower. The shifting of air assets from Ground Forces to the Air Force enforced central guidance of airpower, which in turn reinforced its effectiveness. Irregular warfare in Chechnya showed that lack or absence of expensive PGMs, high tech communications, navigation and targeting systems, as well as all weather and day/night capabilities, limited the effectiveness of airpower. But in spite of the financial problems, Russian airpower demonstrated that it was capable of enhancing its effectiveness without additional financial support, especially by innovations in command and control and by tactical improvements.

Lessons/ Recommendations for Indian Armed Forces

A ground force destined for combat in the high mountains must be tailored to meet the demands of the environment. Logistic support is necessarily difficult and requires more assets than in other less strenuous environments. The force requirement increases accordingly, to both secure and man supply lines and other essential assets, such as artillery batteries. Similarly, decisive manoeuvre in the mountains requires a significant infantry force capable of operating in small units. The force must be unencumbered by heavy loads, and capable of traversing the world's most inaccessible terrain. The full range of firepower, delivered from the air and ground, is necessary to provide overwhelming lethality to the force engaged in combat at high altitude. Aerial munitions are part of the full spectrum of echeloned firepower that should be available to ground forces. Attack helicopters can provide responsive firepower if pilots are trained to fly in thin air and employ "running gunfire" techniques with PGMs and standoff weapons; this is how the IAF trains.

Yet air power cannot be relied upon as the sole provider of the responsive, concentrated fire needed to support ground manoeuvre. Suppressive fire, created by a heavy volume of continuous fire over a wide area, is a necessary complement to ground manoeuvre, and is best provided by artillery. Artillery must be available to forces engaged in ground combat, despite the challenges posed by the high mountains. The Indian Army at Kargil demonstrated the overwhelming lethality of artillery. All weather, responsive fire is essential to manoeuvre warfare on any battlefield, including the high mountains. British and U.S. forces that deployed to Afghanistan after Operation Anaconda brought 105 mm artillery batteries along. They successfully transported these pieces by helicopters throughout the country, proving that artillery can be mobile in the mountains. This had been done by the IAF in 1999 in Kargil.

Low-intensity and short-duration highly confined but fast and intense conflicts are more likely than wider conventional wars. Airpower integrated with the boots-on-ground will ensure quick response to such contingencies. Keeping in mind the resource constraints infrastructure and cumulative skill level, the IAF is best placed to operate the large inventory of light, medium and heavy types that is required in the future. At the same time, greater jointness in operations is an imperative. A strong justification exists for quantitative and qualitative increase in light, light combat, medium and heavy lift helicopters in the IAF inventory; this is already planned and is already happening. It needs to be borne in mind that helicopters inherently are capable of multitasking and could serve, as they already do, in border infrastructure developments and disaster management. In terms of pilot training, there is a need to look for intense but low cost options like simulators, enhanced night capability and special skills which improves the ability of aircrew to adapt to new and changing situations. A core group of pilots should always be available to carry out the more demanding situations of special operations. The IAF's training and recruitment are in sync with the acquisition program that looks at multifarious capabilities well into the future.

The indigenisation programme based on the ALH model has been given a boost. No country, as yet, has a helicopter tailor made for the high Himalayas. Similarly, in the future the MLH and HLH requirement need to be addressed by an indigenous programme. The LCH, as promised by HAL brochures, is an absolute and immediate requirement which needs to be met at the earliest. Modernisation of the helicopter fleet, including night and all-weather capabilities is being taken up as an urgent imperative. Out-of-Country contingencies would best be catered to by incorporating air-to-air refuelling in all future MLH/HLH inductions. Similarly, incorporating PGM capabilities on most helicopter types would act as a tremendous force multiplier. This too is being catered by the IAF.

Scenarios of LICO encompassing counter-insurgency, counter-terrorism or even localised wars like Kargil might evolve in the future in J&K, Nepal, Bhutan, Myanmar and NE India. Some of these could be fanned by adversaries and competitors like Pakistan and China. Highly mobile light infantry and Special Forces adequately supported by fixed wing and helicopters may help contain them from growing into more serious confrontation. In fact, given the scale of involvement and risks that would be associated with helicopter operations in future conflicts in the Himalayas,

combat support is a misnomer-it is very much part of combat operations. Lessons drawn from recent campaigns have forced *a revamp of Man, Machine and Tactics*, which is well underway in the IAF's helicopter fleet.

CHAPTER III

1962: War with China

Introduction While the Sino-Indian War of 1962 mainly evokes strong debates on the health of politico-military structures at that point of time, two highlights of the campaign stand out quite visibly. The first was the bravery and heroism of the Indian Army Jawan where he stood his ground, such as in the Battle of Walong. The other is the herculean and equally heroic exploits of the transport and helicopter fleet of the IAF. In an era of no roads and infrastructure, it would have been a bigger disaster and tragedy had the gallant IAF pilots not risen to the occasion. One must remember that Mi-4s and Chetaks had only been inducted recently. Without time for conversion and operational training, as is taken for granted today when we raise new units, units went in with just 'josh' and a strong empathy for our comrade in arms – the Indian Army. Risk-taking was of the highest levels, sometimes bordering on the suicidal, but the pay-offs in terms of survival of our troops were greater. While the debate on why offensive air power was not used will continue for a long time, the unsung story of valiant and bold helicopter pilots of the IAF needs to be taken note of.

The Soviet Mi-4s were inducted only in Nov 1959. 104 HU converted to Mi-4s while 109 HU started being raised only the same month. 105 HU converted to Mi-4s just before the Chinese attack. The units had no time or luxury for achieving graduated operational status. Aircrew just innovated and learnt high altitude flying by trial and error. Constantly, the envelopes of the newly acquired machines were enlarged driven by operational necessities. Even when pilots and aircraft were downed in battle, aircrew

showed remarkable courage and physical stamina to evade the enemy and undertake tortuous treks back to own areas – so that they could get back into the cockpit to fly again. The professionalism, innovativeness and 'creativity' of the maintenance crowd were evident from the quick recovery of the downed Mi-4 at Walong – it had been given up as a lost cause. The never-say-die spirit of the IAF air warriors of the '62' era would always remain a beacon for the helicopter fleet of the IAF and the Jawans of yesteryears who were witness to their deeds.

The Narrative   The Sino-Indian conflict of 1962 was a war that gradually escalated from border skirmishes to a full blown conflict. The Indian intention right up to 20th October 1962 was to limit its scope and intensity. The role of Indian Air Force in this first bloody conflict between two Asian giants was a limited one; however, helicopter and transport air support proved crucial, as the whole success of 'the forward posture' depended upon it. The role of helicopters in the conflict was critical due paucity of road communications; the deployment, maintenance and survival of ground forces was mostly dependent on air supply. In August 1962, Leh was still not connected by a road. In Arunachal Pradesh (erstwhile NEFA), the build-up of forces on Thagla Ridge area and deployment of Assam Rifles posts in Op Onkar and Leghorn was conditional on logistic support by the transport fleet of the air force and Kalinga Airways. Transport air support was on theatre basis, with the Air HQ only controlling allotment of resources. In the west, No 1 Operational Group was in overall control, and was located at Palam, New Delhi. In addition, AOC J&K and the Tactical Air Centre attached with XV Corps at Srinagar also coordinated the effort. Dakotas and IL-14s were based at Srinagar from where they flew mainly to Leh. Packets were based at Pathankot and Jammu, and flew mainly to Chushul. The AN-12 Sqn was based at Chandigarh and supplied Leh and Chushul. Army HQ laid down priorities and most decisions were taken in joint conferences held regularly. In the East, HQ Eastern Air Command coordinated the effort. The Packets were based at Tezpur while Dakotas operated from Guwahati. Some Packets were also based at Jorhat. The Otters were mainly used to supply the Walong brigade and army posts in Siang, Subansari and Lohit Sectors. Kalinga Airways continued to supply Assam Rifles posts. In addition, IAF Dakotas had been carrying out supply drops to posts in Nagaland, where nearly two brigades were deployed at numerous posts to fight Naga insurgents.

Air supply and casualty evacuation received a boost when 107 Helicopter Unit, equipped with Russian made Mi-4 helicopters, was moved to Leh on 13th May 1961. It could carry 4-6 passengers and was capable of high altitude flying. After the Chinese surrounded Galwan post on 4th July 1962, it was exclusively maintained by Mi-4 helicopters. A record was created when in October 1962; right under the nose of the Chinese, the 1/8 GR Company was replaced at Galwan by a company of 5 Jat. A Squadron of Otters (59 Squadron) as well as 105 and 110 HU carried out daring landings on unprepared surfaces in NEFA to deliver supplies to the army. They inducted an entire brigade (11 Infantry Brigade) from Tezu to Walong. The Squadron which was mainly operating from Jorhat moved a detachment of two Otters to Tezu on 28th September 1962. During the conflict, the Otters flew 982 hours, and air-lifted 414 tonnes of supplies and 2083 troops. The aircraft on return journey from Walong evacuated casualties, saving many valuable lives. This contributed significantly to the raising of morale of the Walong garrison. 105 HU with just three Bells, two Sikorsky S-55s and two Allouette helicopters flew extensively for dropping supplies and rescuing personnel.

105 HU was operating in the Namka Chu area on 20th October 1962. Sqn Ldr AS Williams took his helicopter to Tsangdhar as there was no news from Sqn Ldr Sehgal who had gone in an earlier sortie. On seeing the enemy in occupation of Tsangdhar feature, Williams turned back but his helicopter was shot at by the Chinese. He managed to force land close to Zimithang and was rescued by a Mi-4 helicopter. His helicopter had to be written off. In the meanwhile, the third helicopter at Zimithang was lost when on return from a sortie, it came under Chinese fire. In the chaos of withdrawal, the aircrew and ground staff had to make their way back on foot for three days. 105 HU thus suffered a loss of three valuable helicopters. In November 1962, the unit received two freshly arrived Allouette III helicopters and continued to support the Se La and Dirang garrisons.

Helicopter pilots showed exemplary courage in undertaking 'impossible' tasks, often landing at night with the help of a mere torch light on difficult mountain tops to rescue the wounded. 110 HU was raised only in September 1962, but performed equally heroically. This unit operated the bigger and better Mi-4 helicopter. In a short time, the unit had attained a fully combat-ready status through intensive training. It mainly operated from Tezpur and carried out sorties in the Tawang Sector. A detachment of three helicopters was sent to Walong Sector on 26th October 1962. These machines, along with Otters, helped in the army build up at Walong. As a result of day and night efforts in October and November, an average of 3 sorties per day per pilot allowed 16000 lbs of load to be lifted daily. On 22nd October, they flew 62 sorties from Tawang and rescued 176 women and children from the clutches of the invaders. Once the withdrawal of ground troops commenced, helicopters dropped rations for withdrawing columns and picked up the wounded. Many a soldier owes his life to the brave and untiring helicopter pilots of the IAF. During the supply missions, one helicopter was shot down in Walong on 16th November. The crew managed to escape along with the army. 110 HU carried 180,000 lbs of supplies, 1700 personnel and flew a record 650 hrs: indeed a proud record.

105 HU Diary Records: 30 Dec 1962

Since the commencement of recent hostilities with China this unit has done 716 sorties on various commitments like causality evacuation, reconnaissance, supply dropping, positioning of arms, ammunition and allied stores at various outposts in the NEFA. 558.15 hours of flying was done by Bell 47G-3, Sikorsky S-55 and Alouette III helicopters. 14,625 Kg of load was carried from the rear to forward posts and 356 casualties were evacuated from most of the battle fronts. 135 personnel of army and IAF were airlifted for various purposes.

September 1962   In the month, the unit did not do any appreciable work for the operations as there were only two pilots and one serviceable airlift aircraft available till the end of the month. Then, by the last week of the month, three more aircraft were allotted in and three pilots were attached to the unit. Demand for the unit's helicopter was also not much as such only 16 sorties were flown for the operations.

October 1962   During the month of October, the unit carried out sorties mainly at Tawang, Zimithang, Shakti, Tsangdar and Bumla areas. A total of 293 sorties and 135.05 hours were flown in this month. On 9th October 1962, a detachment of three aircraft along with four pilots and nine ground personnel was positioned at Zimithang for carrying causality evacuation, supply drop, recce and other commitments for 4 Div deployed in this sector. The unit did 270 sorties and 57 casualties were evacuated from battle areas and many other posts along the border where fighting was going on. 4200 kg of load was carried for the army in this sector. In this particular sector and operation alone, the unit has suffered great losses. At Tsangdar on 20th October 1962, Sqn Ldr VK Sahgal, a veteran pilot with many hours of flying experience on different types of helicopters, was taken prisoner by the enemy forces and his helicopter was also captured. He was sent by the local Army Commander at Zimithang to find out the news from Tsangdar as there was no contact between them. Army intelligence was not able to give any accurate information regarding the enemy position. The pilot was left all by himself to look for enemy movement.

The same day Sqn Ldr AS Williams, another highly experienced pilot on helicopter, was asked to go and look for Sqn Ldr Sahgal as he was overdue to return from his mission. Sqn Ldr AS Williams was shot at by the enemy forces when he decided to turn back from Tsangdar after seeing the other helicopter neatly parked, and got suspicious. He was shot at when he was trying to go round from his approach to land, after seeing the enemy movement on the helipad. Luckily, he managed to get out of the area with his crippled aircraft and managed to force land the helicopter on the safer side of the ridge from where he ran back to Zimithang. But for his skill and presence of mind, the IAF would have lost more aircrew and another aircraft to the Chinese. By then the enemy had advanced very fast and our army had no time to warn each and every post or DZ. All the personnel of the detachment were told to withdraw quickly. Sqn Ldr AS Williams who was injured by the enemy fire and other ground crew were airlifted by Mi-4 helicopter from a nearby helipad. But misfortune fell on one officer and eight airmen of the detachment who had been away from Zimithang to salvage the helicopter shot down by the Chinese. When they returned to Zimithang the place was already under enemy fire. They had to run away leaving all their belongings; and walked nearly eight days through the rugged mountainous terrain and jungle without food or shelter. Though they came across many army camps and posts on the way the army authorities did not give them food or shelter even though they recognised airmen and the officer. The unit experienced extreme hardships and suffered a great loss during the period. In spite of all the shortcomings, it had carried out all the task entrusted to it exceptionally well. Zimithang was abandoned with a sad heart as three valuable aircraft were lost and one very experienced pilot taken prisoner by the enemy. Tawang fell to the enemy and there was silence all along the border in NEFA. Fresh threats were encountered in the eastern sector (Walong area) of NEFA in the last few days of October. Demand for recce and casevac sorties were received and these were carried out.

November 1962   Early in the month, the unit received two Allouette III helicopters (French) and trained pilots were also posted in at the same time. By then helicopter operations were exclusively controlled by the AOC No 1 Group to avoid further mishaps like that of Zimithang. In the period of 1st to 10th November, there was a lull in both sectors of NEFA as the Indian Army had withdrawn to Jang and Sela in the West and regrouped at Walong in the East. The unit operated two Bell 47s and two Allouettes for various missions. Slight difficulties were experienced due to the fact that Allouettes had come new in this area, but this was overcome in a couple of days. As days passed, the demand for helicopters increased considerably. Both

Allouettes were positioned at Tezpur and one Bell at Teju to carry out recce sorties for army commanders and generals. At this point, the unit had on its strength, two Bells, two Allouettes and one Sikorsky S-55 helicopter with only a few hours to go for major overhaul. A few sorties however were done by this helicopter to evacuate casualties from other frontier divisions of NEFA.

After 15th November, when the enemy had advanced dangerously close to Walong, there had been frantic demands for helicopters. But on November 16th, Walong had to be abandoned. All helicopters were used to induct fresh reinforcement of troops from Teju to Hayuliyang. It did a few sorties of recce and food dropping for retreating soldiers and casualty evacuation. All pilots did remarkable jobs while carrying out evacuation of casualties. They carried as much as ten passengers in the aircraft from places which were inaccessible to any other type of aircraft or flying machine. At times, pilots landed in places where there was clearance just enough for the rotor to fit-in. This required great skill and involved risk. All these were done as pilots realised the gravity of the situation. Every life saved and every casualty picked up from the front lines contributed greatly towards the morale of the army and the IAF. In this sector, one Mi-4 was shot at by the enemy that was following our retreating army. The crew of the damaged aircraft was picked up by one of the Allouette helicopters operating in the vicinity. The Mi-4 with two crew members had several casualties on board. All of them were saved just because there was a helicopter nearby, otherwise all would have to walk or run all the way back. Above all, the IAF would have lost the services of two more valuable helicopter pilots at this precious time. This incident indicates that on all operations where helicopters are required to carry out commitments in the vicinity of the enemy movements, over thickly wooded foot hills, valleys and ranges of the NEFA, it is essential that they must operate in pairs.

On the 18th November, the unit was asked to send one Bell helicopter to Dirong Dzong directly from Jorhat. Sqn Ldr AS Williams was the captain and his Bell helicopter did not have enough endurance for the complete trip from Jorhat to Dirong Dzong. He had to land at Tezpur for refuelling and it was a 'Blessing in Disguise'. Dirong Dzong had fallen by then and Bomdilla was under enemy fire on the same day. There was no contact between the Army Division at Dirong Dzong and the Corps HQ at Tezpur. All flying was stopped for the day. The enemy had run over Sela, Senghe, Dirong Dzong, and Bomdilla was being shelled. That was a great shock to all the aircrew that were operating in this theatre, when they realised that all their efforts – flying done by day and night, to drop supplies and airlift personnel to the forward posts – had all gone waste. A gigantic task was given to them and they did it, but they had to see their efforts going waste. On the 19th November, the unit was entrusted with a task of searching for the retreating stragglers. There was no proper briefing on the possible routes of retreat. It was more or less a suicidal mission to fly over a territory which was not protected by our army and with no information about enemy movements in that region.

Two Allouettes and one Mi-4 took off from Misamari with 300 kg of cargo rations in each and carried out recce of the foot hills towards west and south-west of Bomdilla. They landed at a place called Kalaktang where they spotted a few army jawans marching back towards the plains. The sortie was successful and all helicopters dropped their load and brought back casualties. This particular mission was the beginning of an extraordinary operation by IAF helicopters. On the 20th there was no flying as a general evacuation from the northern plains of Brahmaputra River was ordered. On 21st November, after the announcement of cease fire by the Chinese, plans were made to carry out mercy mission for the stragglers of entire 4 Div from this sector. Both Allouettes were to operate from Amtulla at the foot of the hills in the Kameng FD of NEFA. All the helicopters were used for this mission and all were marked with a Red Cross. In fact, helicopters

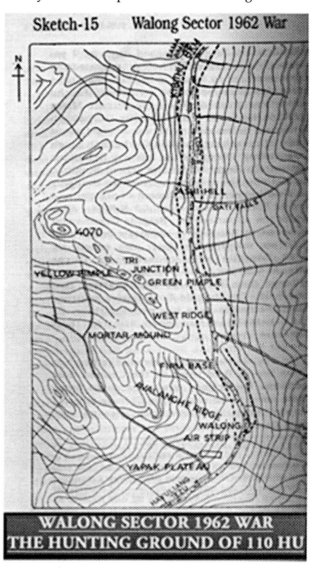

Sketch-15   Walong Sector 1962 War

WALONG SECTOR 1962 WAR
THE HUNTING GROUND OF 110 HU

were doing the job of an airborne ambulance. The operation carried on till the end of the month. Pilots encountered peculiar hazards. On many occasions exemplary skill and keenness was shown by the pilots. In one day, as much as 7 to 8 hours of flying was done and each day helicopters dropped nearly 1000 kg of food and compo ration, and evacuated 35 army personnel and casualties from the dangerous jungles and narrow valleys. Sometimes, pilots had to fly deep in very narrow valleys to search for starving stragglers.

In one particular mission alone, more than 100 sorties were flown and 7000 kg of load dropped for the Army. More than a hundred casualties were evacuated and the top ranking officers of the Division including the GOC, Maj Gen Pathania MC, were evacuated from the jungles after having walked for four days. Many lives were saved and many starving Jawans were given hope and fresh strength. Their low morale was boosted up by these operations. But for the performing ability of Allouette III and will of the pilots flying it, this operation could not have achieved the great success it did. The unit carried out all the task entrusted to it. There were only a few aborted sorties due to bad weather or reasons which could not be avoided in any way. Though general peace prevailed in the border area and fighting had stopped all over, it still carried on mercy missions till the end of December 1962. More than 90 sorties were done in the month of December to give food to the stragglers coming back through all the divisions of NEFA and nearly 70 casualties were evacuated.

Through out the operation, morale of the unit remained very high in spite of the drawbacks and misfortune. All the aircrew and ground crew had to undergo extreme hardship on many occasions due to frequent changes in the plan of operation or briefing for commitments. It was seen that Allouette III helicopter proved extremely useful for all type of mission in the hills and jungles of NEFA and Assam. Bell 47G-3 helicopter was used for almost all type of commitments in spite of its short range and endurance. The one S-55 helicopter held on the strength of the unit could not be used for the operations except for extreme emergency cases of requirement from other place places of NEFA and Nagaland. During this operation, OC of the unit, Sqn Ldr AS Williams, was awarded Vir Chakra for the gallant work he did by flying the Bell 47G helicopter at night. He used a hand torch light as the helicopter was not installed with any night flying kit. He flew the aircraft at night over the hills and mountains with this light to drop ammunition and to bring back casualties from the battle area.

110 HU Records: Vanguards

The pioneer team of 110 HU got together in Jul-Aug 1962. After raising the unit, some pilots who had converted on to the Mi-4 helicopter in USSR took on the mantle of converting other pilots. In a short time, the unit had attained a fully combat ready status through intensive training. Within two months of its formation in September 1962, the Unit answered the 'Clarion Call' and took active part in the 1962 Ops at Walong and Tawang sector. When the bubble burst on 20th October, the demand for helicopters from 110 HU increased tremendously and pilots flew round the clock on varied missions. A detachment of three helicopters was sent to Walong Sector on 26th October. The unit helped the Otters in the army build up at Walong. As a result of day and night efforts in October and November, the unit carried out an average of 3 sorties per day per pilot and lifted 16000 lbs of load daily. In October 1962 alone, the helicopters flew 62 sorties from Tawang and rescued 176 women and children from the clutches of the Chinese troops. Once the withdrawal of ground troops commenced, the helicopters dropped rations for withdrawing columns and picked up the wounded. The Unit during the entire operations carried 180000 lbs of supplies, 1700 personnel and flew 650 hours, a record in which any unit can take justifiable pride. Many an Army man during these operations owes his life to the untiring and heroic efforts of the 'Vanguards'. Flt Lt KK Saini, the OC was awarded the Vir Chakra for his courage and determination displayed during a rescue mission in 1962 operations. On 16th Dec, he handed over command to Flt Lt Banerjee.

Personal Account of the Technical Officer 110 HU
(Air Cmde Melville Rego in Bharat Rakshak)

It was Winston Churchill who said after the Battle of Britain during the Second World War: "Never in the field of human conflict was so much owed by so many to so few." In the context of the Chinese War of 1962, in the NEFA, Ladakh, Leh and Bhutan sectors, I would like to modify the above to say: "*Never before in the history of Indian Air Force conflicts have so few risked their lives as 'sitting ducks', under enemy fire, in primitive unarmed helicopters, to save the lives of so many.*" Many records were broken in terms of tonnage of supplies dropped, casualties evacuated, lives saved, flying hours, landings to deliver supplies, and others. The importance of Tezpur airfield at its strategic location in Assam, 20 miles from foothills on the NEFA border, must have dawned suddenly at the top level. Capture of the airfield (with its defenders – IAF ground staff of 11 Wing & 110HU) would have thrown open the whole of Assam with its oilfields and more. This coupled with the reverses suffered by the Army in NEFA, prompted the Govt to set up the high level No 1 Operational Group and 5 TAC at Tezpur in mid October to conduct Air Force activities in conjunction with HQ 4 Corps. 11 Wing and 110HU now came under the functional control of No 1 Group. A Group Liasion Officer (Major) was positioned at our base with a LMG and a few jawans to protect the airfield. If war was to be won by Blood and Guts alone – as Gen. George Patton said, was this back-up sufficient? A major decision at this time was to order our fighters and bombers NOT to take any part in the operations and consequently the Vampire and Toofani squadrons were flown out to bases in the rear, leaving 110 HU to face the music.

The situation in the Subansiri, Lohit and Kameng sectors of NEFA had become hopeless with the fall of important bastions, but the Army still felt it could hold on to the Se La Pass. When this was cleverly by-passed by the Chinese, the threats to Walong and Tawang–Bomdi La road increased. As winter had set in and weather conditions were hazardous, airdropping of supplies could not be carried out effectively, but sorties were done to evacuate casualties, frost-bite cases etc. In retrospect, while our transport fleet of Dakotas, Packets, An-12s, Il-14s and Super Connies did a magnificent job in their spheres of action, the role played by our Mi-4s in the mountains of NEFA was no less praiseworthy.

The Incident at Walong   The battles at Walong constitute perhaps one of the greatest acts of heroism in Indian Army history, where about 3000 officers and men repulsed no less than 15 fierce attacks by a full Chinese division of 15000 men. Eventually, Chinese firepower, manpower and superior tactics prevailed and our forces were overwhelmed, suffering many casualties. Many cases of bravery and heroism of our officers and jawans came to light later. On 18th Nov 62, our OC, Flt Lt K K Saini and Flt Lt K K Deb were ordered to evacuate battle casualties from the Walong area by landing at a helipad, reported to be free of Chinese troops. As they approached the helipad, the Chinese opened fire and the helicopter was hit at several places causing severe damage. Flt Lt Saini himself was badly injured in the leg and started bleeding. He used great presence of mind and steered the helicopter to safety to avoid further damage by enemy fire thus saving the helicopter and the life of his co-pilot. Flt Lt Saini was awarded the Vir Chakra for his courage, determination and professional skill. This was one of the few VrCs awarded to Mi-4 helicopter pilots.

Enter the Dragon   News had got around that the Chinese had cut the road near Bomdi La and were heading southwards. On 18th Nov, the airfield and surroundings was a beehive of activity. Daks were landing and taking off at regular intervals. There were 15 of them, parked nose-to-tail on the tarmac. Our Station Commander, Gp Capt Bhawnani was all over the place, calm and cool, stoic and giving instructions. Can you believe that he used to personally marshal the aircraft after landing, and guide them to their parking spots, for he knew their schedules exactly! Evacuation of civilians went on in full swing. Our helicopters were dispersed to nearby army helipads where they could be of use when required. One was left at base to make a Saigon-like escape! Tezpur was a ghost town that day, deserted and all shops locked up. At the officers' mess, all cooks, waiters, sweepers, our personal bearers and other civilians had run away. The Home Ministry had, I believe, asked its entire staff at Tezpur, to move out and they did so eagerly. The DC of Tezpur also pushed off, for which he was later chastised. At other places, they were to burn the oilfields and power-plants. The State Bank of India made a big bonfire of all its banknotes and shut up shop. Our families back home were desperate for news, but there was no means of getting in touch with them. Many had given up hope of our survival. The Army GLO, his staff, the LMG and about 50 men, took off leaving the airfield defenceless. We were told to place kerosene oil tins, axes etc at all vital installations, in readiness for destruction at short notice. In deference to our Service personnel, I must say they stuck to their posts loyally. Our admin and logistics staff was ordered not to stick to procedures. No vouchers, for instance when items were required urgently – just a note.

From early morning, there was a procession of cars landing up at our tarmac. The occupants turned out to be tea planters with their families. They rushed to us, after locking their cars, saying that they wanted to go to various places – Calcutta, Bagdogra, and even UK, many even offering the car keys to us. "Boy, we're off, you can have my car" (I wish I had kept the 3 car keys offered to me)! Here again, our Stn Cdr was in full control of the situation. He had detailed an officer to make note of all the cars around, the names of the owners, registration numbers and even the possessions left behind in the car (as far as possible). Then he directed the men to take their families to a particular Dakota depending on their destination. The Dakotas flew off one by one. It may be hard to believe, but there were 53 cars parked at the airfield that day. Around 25th Nov, when the whole show was over, they all returned, asking very sheepishly "May I have my car please" To the credit of the Air Force and its solicitude, every single car was returned to the owners, as per the list. It was gratitude unlimited thereafter between the Planters and the IAF at Tezpur.

Misery at Midnight – The Last Supper   The night of 19th Nov was perhaps the most traumatic of my life. The Chinese had finally arrived at the foothills, just 20 miles from our airbase. Storming the airfield was imminent. Early evening, we were told to go to our pilots rooms (who were not at base), pack one box each of their personal belongings and bring it to the tarmac. All personnel were instructed likewise. This gave us the impression that perhaps the plan was to fly all of us out – about 200 officers and airmen from the Wing and 110HU, after setting fire to and destroying equipment. But certainly there were no aircraft for this. I believe there was a plan also to destroy the airfield runway, but then how would they fly out? The tension of waiting to be massacred is definitely worse than going boldly into battle and riding your luck. You see movies or read novels of suspense, but this was the real thing. Gloom and despair all around – We thought of our families – wives, children (my son, just eight months old may not see his father), went about bolstering our men's sagging spirits and wondered what was going to happen to us in the next 12 hours. Then a call came at around 10pm for all officers at the airfield to assemble at the Mess.

Our Stn Cdr was at the head of the table and when all were seated, 'Dinner' was served– just bread and onions, because the cooks and waiters had vanished. Well, we thought, this is going to be our 'Last Supper', so let's enjoy! Then he started talking and his words ring in my ears to this day: "Boys" he said "This is your last night alive. I can't do anything more for you. The Chinese are only 20 miles away, thousands of them and they can cover this in one night. Expect them early tomorrow morning. We are on our own and there is nobody to defend us. Now go to your rooms, say your prayers and hope for the best. On the way, go to the armoury; collect a revolver and six rounds. When the Chinese come, kill six of them, and then put up your hands. Thank you all for your support. So good night." We all trooped out, quite distraught, no questions asked, knowing he must be at the end of his tether (putting it mildly) and slinked back to our rooms, via the armoury, with the revolver and ammo. No question of sleep and I just tossed in the bed, restless.

Cease-fire   Early on the morning of 20th Nov, all of us were lined up on the tarmac, with our baggage, wondering what happened to the Chinese. No reports had been received about any advance into the Assam region, including Tezpur, which was the reason why there was no setting fire to or destruction of our equipment. Suddenly, at 7.35 AM, one of my airmen holding one of those quaint transistors to his ears, gave a big shout "Sir, BBC has announced that the Chinese have declared a unilateral cease-fire." We could not believe our ears, but confirmation came shortly from official sources. We just slumped to the ground. God's ways are truly inscrutable! The other bit of good news we got was that both the pilots, who had been fired upon, were safe. It was only by 24th Nov that we were assured that there was no more fighting and the Chinese were withdrawing. The tasks of 110 HU were not over, however. Supply and rescue missions, had to be carried out further, as many of our troops were scattered over the hills. My own cousin was one of the stragglers.

Servicing and Salvage of the Helicopter at Walong   A few days after the ceasefire, we got news that the helicopter which was shot down was lying near Walong and appeared to be badly damaged. After getting clearance that the area was clear of Chinese, we mustered a group of officers and technicians and set off in two helicopters, carrying necessary tools, spares and salvage equipment. Everyone thought that we would just be able to salvage the removable equipment and return. When we saw the helicopter, BZ 587, (I still remember the number), the heavy damage was clearly evident: the oleo legs had been punctured and hydraulic oil had leaked out, an explosive device had been planted in the cockpit and the instruments blasted, electrical and signal cables had been cut at the junction, the VHF R/T crystals had been removed, and the right side of the helicopter was riddled with bullet holes. Why not patch it up, so that it could fly with minimum instruments, assuming the engine being OK? All the concerned tradesmen were there, so under the expert supervision of our EO Flt Lt Jain, the servicing and repair was carried out over a period of two weeks. The nearest army unit helped with rations and materials necessary for the job. On the hills around, the dead bodies of our brave Jawans of many regiments and communities could be seen. Flt Lt Jain was given the privilege of flying the helicopter back to base, (with some coaching from the trained co-pilot) after it was declared airworthy by the pilots.

The scene at Tezpur airfield, as described to us, was something out of this world. The helicopters approached in a formation with BZ 587 in the lead, peeled off in turn and then landed to tremendous cheers from all the personnel on the tarmac. I'm sure our present Sarang crowd would have been amazed at the heli-batics! Unfortunately, no cameras, cell phones, webcams etc were available to record the event. The reverse flow to Tezpur started with the Planters, POWs staging through in Daks and civilian aircraft. Normal life soon built up. An American C-130 landed and stayed for an hour. I understand that it was on loan to us since Nov 1962 after our Govt had appealed to the U.S authorities for supply of squadrons of fighter and bomber aircraft.

Return of the Tribal Chief   The Tribal Chief and Head of the monastery at Tawang had taken refuge in India in mid-October. Early December, we were told to drop him and his retinue back. We were not sure if Tawang was clear of the Chinese and even if the troops had withdrawn, some spies in disguise may still be around. One factor in favour was that the tribals in that area were Buddhists and totally averse to the Chinese especially after the mauling of Tibet. After getting final clearance, Debu and I took off for Tawang. As we approached, we saw crowds near the helipad then they started waving. We then decided to risk a landing. The scene after we landed was like Eden Gardens after Tendulkar had scored a century, or India had won a cricket match. They fell at the feet of the Chief and embraced us.

REMEMBERING THE BRAVE:
GALLANTRY AWARD CITATIONS

Vir Chakra

Squadron Leader Surya Kant Badhwar (3973), General Duties (Pilot)

Flight Lieutenant Kuppuswamy Lakshmi Narayanan (5053), General Duties (Pilot)

One of our Army posts in Ladakh on being surrounded by Chinese invaders had to depend solely on air support. Our troops, who had endured the siege, had to be relieved and this could be done only by helicopter. On 4th October 1962, Squadron Leader Badhwar and Flight Lieutenant Narayanan flew several sorties to carry out this task. During one of the sorties, they experienced loss of engine power and eventually force landed in a river bed after avoiding a few Chinese posts. In order to obtain relief, they walked towards one of our posts, but after some time they observed a large number of armed Chinese troops. They then returned to the helicopter and decided to chance a flight. Meanwhile, about 150 Chinese had taken up position on hill tops near the helicopter and were beckoning to them to surrender. With great presence of mind, they managed to get to the helicopter and piloted it safely to one of our posts and thus saved valuable lives and equipment.

Squadron Leader Arnold Sochindranath Williams (3950), General Duties (Pilot)

On 20th October 1962, Squadron Leader AS Williams was ordered to find out the position prevailing at one of our Army posts in NEFA which had been subjected to heavy attacks by Chinese invaders. To carry out the task, he flew in a helicopter but it was damaged by heavy anti-aircraft fire by the Chinese. In order to save the aircraft from further damage, he skilfully manoeuvred it towards the nearest helipad at another post. Unfortunately, the helicopter developed engine troubles as a result of the damage caused by enemy fire and he was forced to crash-land it two miles from the helipad.

He then trekked to the nearest Army post and was eventually evacuated to his own unit. Earlier, on 12th October, he had flown several sorties in a helicopter by day and night to evacuate casualties from forward posts in the same area.

Squadron Leader Chandan Singh, (3460), General Duties (Pilot)

On 20th October, 1962, Squadron Leader Chandan Singh was detailed to carry out supply dropping in the Chipchap area in Ladakh. On reaching the dropping zone, he noticed that the outposts were under heavy fire from the Chinese forces. He successfully dropped vital supplies to our garrison; although his aircraft was hit 19 times by enemy ground fire. He displayed courage and devotion to duty in carrying out the task in complete disregard of his personal safety.

Flight Lieutenant Krishan Kant Saini, (4436), General Duties (Pilot)

Flight Lieutenant Krishan Kant Saini had been operating in NEFA area since October 1960. On 18th November, 1962, he, along with his co-pilot, was evacuating seriously injured battle casualties in Walong area. He was instructed to land at a helipad close to enemy line which was reported to be clear of enemy troops. When he was over the helipad, Chinese troops opened fire from many directions. His helicopter was hit at several places: the main reductor was damaged and oil gushed out from it in a thick spray which blinded him temporarily. His right ankle was also injured by splinter and he was bleeding profusely. With great determination, presence of mind and skill, he nose-dived the helicopter almost to ground level in order to avoid further damage from enemy fire. He thus saved the helicopter and the lives of his co-pilot and passengers. In spite of the damaged hydraulic system and the personal injury, he skilfully brought the aircraft back to base.

A TRIBUTE TO THE FIRST MARTYR

SQN LDR VINOD KUMAR SAHGAL VM (3853) F (P)

Date of Casualty	:	*20th Oct 1962*
Unit	:	*105 HU*
Type of Aircraft	:	*Bell 47 G-3 (BZ 543)*

The IAF was not called upon to participate in the 1962 Sino-Indian conflict except to provide transport and logistic support to Indian troops. Sqn Ldr VK Sahgal was part of a detachment of 105 HU based at Lumpo in the erstwhile NEFA (Arunachal Pradesh). The detachment was operating three Bell-47G helicopter, which were flown single pilot and capable of carrying two passengers. The detachment which began operating on 9th Oct 62 with 4 pilots and 9 ground crew personnel, carried out casualty evacuation, supply drops, recce and other commitments for No. 4 Div employed in the sector.

The Chinese attacked all the Indian Army Posts along the Namka Chu at 0500 hrs on the 20th of October. The independent Mortar Battery at Tsangdhar could not be contacted on R/T as the Chinese had jammed the radio frequency used by the gunners. With all contact at Tsangdhar cut off, Sqn Ldr Sahgal was ordered to fly into Tsangdhar at about 0830 hrs with Maj Ram Singh, 2nd in Command of the Div Signals Regiment. Unaware of the tactical situation, the helicopter proceeded to Tsangdhar (elevation 14,500 feet).

A short-while later a Packet flown by Sqn Ldr Plomer and Flt Lt Sadatulla on a drop sortie to Tsangdhar heard urgent calls from another Bell helicopter flown by Sqn Ldr Williams, CO, 105 HU, trying to raise Sqn Ldr Sahgal on the R/T. After contacting Sqn Ldr Williams on R/T the Packet proceeded to fly overhead Tsangdhar DZ. They saw Sqn Ldr Sahgal's helicopter parked on a small clearing a little distance to the left of the DZ with no one nearby. The entire DZ was covered with white drop parachutes. There was no movement and they could not see a soul anywhere around the area. As they came in for their dropping run however, they saw a few bright orange flashes on the DZ kicking up small pockets of dust between the pile of parachutes and the parked helicopter. The Packet turned back, simultaneously informing Sqn Ldr Williams that the other helicopter was on ground at Tsangdhar with no one near it and that the DZ was being shelled.

The fate of Sqn Ldr VK Sahgal was never known and remained shrouded in mystery, though his passenger appears to have been captured alive. Sqn Ldr Sahgal was listed as 'Missing in Action' w.e.f 20th Oct 1962 and though he was subsequently listed as 'Confirmed Killed' in September 1965, the exact circumstances of his death are not clear.

Rest in Peace

VINTAGE VIGNETTES

Mi-4 Relief Operations in Assam; Late Sixties

Induction of Chetak in 111HU; Marshalling a Mi-4 in Rajasthan

Bringing Succour to our Bretheren: Mi-4 in Bengal and Orissa

Action Stations to Liberate Bangladesh: Indo-Pak War 1971

Epitome of Joint Planning & Execution; Army/Air Force 1971

Rescue Operations 1962 Walong; Winching up in Kutch 1965

The Magnificient men Maintaining their Machines: 1962 NEFA

The Venerable Workhorse (Mi-4) Giving Way to the Powerful Mi-8

Above All, Helicopters Save Lives

CHAPTER IV
The 1965 Indo-Pak War

Introduction Helicopters had hastily been put together as a coherent act during the 1962 debacle. In the intervening years, the IAF had the time to train its helicopter crews in various roles including safe high altitude action. However, the focus was on benign roles such as logistic support and evacuations. But as each new war brings along its fresh surprises and 'lessons-to-be-learnt', the developments in 1965 when Pakistan pushed in hordes of infiltrators into J&K, required a helicopter role that had not been envisaged – the armed offensive support.

The IAF leadership displayed remarkable vision to conceive that only helicopters could operate in narrow valleys searching out loose bands of Razakars and take them out. In a well-crafted plan, the IAF modified the Mi-4s for bombing and gunnery role, fast-tracked aircrew training and deployed the task-force for action– all in record time. Once again personnel displayed copious doses of innovativeness, creativity and a spirit of 'can-do' to take on the challenge. The operations themselves were the riskiest air operations undertaken during the 1965 war. Enemy small arms fire and mortar only added to the challenge of the mountainous terrain. Srinagar, the mother base was under continuous threat and attack of Pakistani fighters. Those were the days of no early warning and effective air defence. On more than one occasion, helicopters were caught and damaged on ground by attacking aircraft. The maintenance crew did a commendable job in putting most of the damaged helicopters back in action. Downed aircrews, just like in 1962, displayed a remarkable sense of survival in trekking through contested and active terrain to escape capture and return to fly again.

IAF helicopters ably supported the Indian Army's onerous task of, first, evicting the infiltrators, and then, take on the Pakistani war machine. While there is some debate on the effectiveness of the Mi-4 in the offensive role, the results on the ground as documented by the Indian Army validates the concept. More importantly, the morale factor of the ground troops seeing air support can never be ignored. This was a key difference in the 1962 and 1965 campaigns, the large number of casualty evacuations from combat zones only added to this critical support to the trooper on ground– all this despite a rudimentary communication system available then. This chapter focuses on the IAF's role in countering the Pakistan launched Op Gibraltar. This penchant for intrusion into J&K has repeatedly been displayed by our neighbour, and has every time been thwarted by the Indian Armed Forces. The IAF's helicopter fleet can take justifiable pride in its legacy of supporting and enabling such operations over the past four decades.

Background to 1965 Ops The fighting in Kutch in the spring of 1965 came before the military balance had titled decisively against Pakistan. The Kutch imbroglio appears to have been unplanned and not the result of any great Pakistani design. The continued inflow of massive American military aid had engaged a mood of overconfidence in the Pakistani armed forces, particularly because the Govt in Pakistan was a military dictatorship under Field Marshal Ayub Khan. So Pakistan threw in its main battle tanks in what was basically a border skirmish between paramilitary patrols of India and Pakistan. The result of the fighting in Kutch was hailed as a military victory in Pakistan. India's failure to hit back in full force, at Kutch or elsewhere along the Pakistan border, was taken to be a proof of timidity and lack of confidence, and Pakistan Army experienced great euphoria. The historic moment seemed to have arrived to solve the Kashmir problem in its favour.

In terms of politico-military interface, our performance was indeed good, due mainly to Lal Bahadur Shastri, the Prime Minister. There was complete understanding and trust between the civil govt and the military establishment. The civil set-up fully backed the armed forces, and the civil population identified itself with the war effort to a degree unseen before. International diplomacy was handled competently, as evidenced in dealing with China, the USA and the USSR. However, inter-services cooperation was far from satisfactory. The institutional framework for it was rudimentary, and the situation on the ground left much to be desired. The Indian Navy had a minimum role in the war. Army-Air cooperation was primitive and initially the IAF was hardly in the loop. Many senior officers had no experience of a modern war and a very inadequate appreciation of the potentialities and limitations of air power. By about 20th September, Pakistan was exhausted with its armoured formations receiving a heavy battering and artillery ammunition over. India would have won a decisive victory if the war had continued much longer, but the Indian Army Chief was overcautious, and 'cease fire' was declared. Tashkent saved Pakistan. The Kutch Operation was, in some sense, a victorious war for Pakistan; but it learnt a wrong lesson that it could walkover in Kashmir. This led to Op Gibraltar and ultimately the September War.

OP GIBRALTAR

Pakistani infiltrators were organized into various task forces and each of these forces was stiffened by inducting officers and some trained troops from the Pakistan Occupied Kashmir (POK) battalions for command and control. The remainder of the personnel came from the Razakars and Mujahids. The Razakars formed the bulk, constituting about 70 per cent of this force. This government-sponsored organization was formed in POK in August 1962. Under this, all able-bodied civilians living near the border were recruited and trained. The Mujahid force was organized in POK much later, and they were used primarily as porters. From the interrogation of the captured prisoners, it appeared that the Mujahids and the Razakars were not volunteers, and most of them had been recruited forcibly under the order of the local civil authorities.

The infiltrators comprised eight to ten 'Forces', each having six units of five companies. Each force was commanded by a Pak Army Major and had been given a code name. Each company was commanded by a Pak Army officer of the rank of captain and below and was known by the name of 'Commander'. A company was made up of one officer, one to three JCOs, half a dozen NCOs and about 35 key personnel from the POK battalions or units of Northern Scouts, 3 to 4 other ranks from the Special Service Group, and about 70 Razakars/ Mujahids, making a total strength of approximately 120 all ranks. The POK soldiers formed the hard core of the companies and the Special Service Group men handled the explosive for carrying out demolition and sabotage. The hard-core personnel had been carefully selected, and the majority of them came from the commando platoons of POK battalions. In certain cases, the entire company consisted of personnel of regular forces, and in other cases the ratio of regulars and irregulars varied from company to company.

Training   Razakars were given training in POK at Nikial, Khuiratta, Darman, Kalargala, Tarkundi, Bohri Mahal, Pir Kalanjar, Rajira, Kotli and Bher by the POK battalions. They were subsequently given intensive training along with the regular troops at various centres. Grouping up between the regular troops and the Razakars was carried out during further six weeks intensive training at the Guerilla Warfare Schools located at Shinkiari, Mang Bajri, Dungi and Sakesar located at Shinkiari, Mang Bajri, Dungi and Sakesar in POK. Training was imparted on: laying of ambushes; destruction of bridges and disruption of lines of communication; raids on military formation headquarters and supply dumps; toughening exercises; and, unarmed combat.

Objectives and Tasks The main Pakistani objectives behind the launching of the infiltrations operations were:

- to establish infiltrator bases for operation at various points within the state of Jammu and Kashmir with the help and support of anti-Indian elements;
- to fan out from the bases as and when possible in order to commit acts of sabotage and violence, to terrorise peaceful and loyal citizens and to provide support to pro-Pakistan elements;
- to attack civil and military personnel and government institutions in different parts of the Jammu and Kashmir Sate so as to disperse the Indian army and the police forces as much as possible;
- to create tension and unrest in Jammu and Kashmir and instigate lawless activities with a view to paralyzing the administration and projecting picture of internal revolt in the State; and
- to facilitate the induction of increased numbers of armed forces from Pakistan.

The tasks assigned were: destruction of bridges and disruption of lines of communication, raids on ammunition dumps and supply dumps; raids on Indian formations and unit headquarters; and, ambushing of convoys and patrols. Having carried out these, they were to merge with the local population and await further orders. They were organized into various forces and operated as follows:

- Tariq Force – Sonamarg, Dras and Kargil areas.
- Qasim Force – Kupwara, Gurez and Bandipur areas.
- Khalid Force – Trehgaon, Chowkibal Nangaon and Tithwal areas.
- Salauddin Force – Uri sector and Srinagar valley
- Ghaznavi Force – Mendhar, Rajauri and Naushahra areas.
- Babar Force – Kalidhar Range and Chhamb areas.
- Murtaza Force – Kel area.
- Jacob Force – Minimarg area.
- Nusrat Force – Tithwal area.

Infiltration Routes

Route	Important Places on Route	Highest Point
Gultari-Drass	Through Gultari, Marpola, Goson, Moradbag on to Drass	4760 mtrs
Buniyal-Drass	Buniyal, Shandorilla, Bhimbet, Drass	5290 mtrs
Matiyal-Kharbu	Matiyal, Chehumudo, Palawar, Kakshar, Kharbu	5090 mtrs
Dusnail-Chunagund	Dusnail, Kirkitchu, Chunagund	2740 mtrs
Brielman-Kargil	Brielman, Yourbal Tak, Simul, Kargil	5380 mtrs
Natsara-Kargil	Natsara, Olthing Thang, Musbar, Chuli Chang, Batalik, Simul and Kargil	2440 mtrs

Extensive infiltration of the Gibraltar Force at various points across the 750-Km long cease-fire line and the international border between Pakistan and Jammu and Kashmir began on 5th August 1965. The activities of the armed infiltrators covered areas from the south-western tip of Jammu and Punch and Uri in the west, Tithwal in the north-west, Gurez in the north, and Kargil in the north-east. Initially, about 1,500 of them crossed the Indian border in Jammu and Kashmir surreptitiously in small batches, and concentrated at selected points inside Kashmir to organize themselves into larger groups. This force was equipped with light, automatic weapons, and its aim, apart from sabotage, was to indoctrinate the Kashmiris so that they could rise in a rebellion against India. Infiltration was mainly directed towards Kanzalwan, Keran, Tithwal, Kargil, Uri, Gulmarg, Mendhar, Punch, Rajauri, Naushahra and Jammu areas.

The second wave of infiltrators was inducted into Jammu and Kashmir in the third week of August. At this point of peak strength, their numbers stood between 5,000 and 6,000. Taking into account the replacements for those who infiltrated, it is estimated that on the whole, a total strength of about 8,000 took part in these operations. By about the first week of September 1965 when open hostilities started between India and Pakistan, a third wave of infiltrators, approximately 5,500 strong, was ready in POK for induction. But this could not be sent across the Indian border due to the operational pressures in West Punjab. The infiltrators initially worked independently in small groups. Subsequently, as worthwhile success was not achieved, they changed their tactics and attempted to concentrate themselves in selected areas and operate in larger groups. Some of the infiltrators from various columns managed to exfiltrate without participating in any operations while other kept drifting and operating indifferently till they ran into another group and merged with it. Towards the later part of their operations they were able to establish their bases and consolidate themselves into strongholds in certain areas. Most of these were in remote, isolated mountainous regions which were not frequented and were not easily accessible.

Helicopter Action against Op Gibraltar   During the dark days of Operation Gibraltar, a helicopter task force, initially consisting of two squadrons but later raised to three, was formed to assist in fighting against the Pak armed infiltrators who had entered Jammu and Kashmir in August 1965. This task force was mainly based in Srinagar, and it carried out 79 offensive sorties against the infiltrators from 20th August 1965, till the end of the hostilities. These IAF helicopters, suitably modified, bombed and strafed the positions of infiltrators in many areas, especially Haji Pir Pass, Tangdhar, Badgam, Mandi, Budil, and the hills around Gurez. While these offensive sorties did inflict much damage on the enemy, more importantly, they certainly exerted a great demoralizing effect on the Pakistani guerrillas. The helicopters also played an important logistical role by dropping approximately 92,000 Kg of essential stores and urgently needed ammunition to army columns operating in different areas, lacking suitable ground communication. They also performed a useful task by speedily evacuating critical casualties from inaccessible areas, flying a total of 198 trips, each loaded to maximum capacity. Some of these helicopters, including three Allouette, were used by senior army officers to get a good view of the areas of operations, so that quick decisions could be taken to plan and execute counter-offensives against the infiltrators. Thus, the Indian Air Force contributed its air effort, limited to helicopter sorties only, till 1st September 1965, when its other aircraft also joined the fray.

Helicopters proved to be especially useful in Jammu and Kashmir for the following tasks:

- Transportation of urgently required defence stores, arms, ammunition and other equipment during critical moments and operations.
- Evacuation of serious casualties from difficult areas, with consequent good effect on morale.
- Special reconnaissance over large areas, especially in sectors where other means of transport were not available.
- Tracking and hunting of enemy infiltrators in terrain almost inaccessible to regular columns.
- Use as Air Observation Posts.

Action at Air Force Station, Srinagar – up to 30th Nov 1965

Mi-4 helicopters were rushed to 1 Wing during the middle of August at a very short notice to give active support to ground forces engaged in operations against the infiltrators. The aircraft had been quickly modified to drop 20 lb bombs and were fitted with 0.5 inch guns. The helicopter detachment consisted of aircraft and crew drawn from 107, 109 and 111 HUs. These started functioning as a detachment with effect from 18th Aug 1965. During the operations, one Chetak from 114 HU was also positioned. In the beginning, on an average there were seven aircraft, which towards the end reduced to four. From Oct 1965 the detachment comprised of 111 HU only.

Helicopters went into action immediately on arrival. In spite of the fact that the aircraft gun had been improvised for the first time for this offensive role and the crew had no previous experience, operations in the entirely new environment and difficult terrain met with ample success. Pakistani infiltrators suffered losses and were demoralised, and our own forces had effective air support in these difficult guerrilla type operations. The entire available helicopter offensive potential could not be fully utilised because the infiltrator's hide-outs could not be singled out from our own inhabitants and 'Bakarwals' (nomads). Further, the lack of quick means of communications and inadequate reliable intelligence sources handicapped offensive operations.

On the outbreak of open hostilities between India and Pakistan, all transport squadrons were moved out of J&K area. This helicopter detachment was called upon to give logistics support to our troops, in addition to its earlier commitment of operations against infiltrators. Thereafter, they started operating even across the cease fire line and just behind the line of actual control. They were engaged in bombing, strafing, casualty evacuation, dropping/ landing of supplies and ammunition, communication and recce operations. Operating from first to last light under the most stressing circumstances and from quickly improvised helipads, they successfully dropped supplies at dropping zones where possibly no other aircraft could drop. At times, some of the helipads and dropping zones (DZs) were within the range of enemy fire, and on some occasions they operated in forward areas where actual shelling was taking place. The mother base at Srinagar airfield was constantly under threat from enemy air attacks. This airfield had neither any early warning facility nor any effective means of air defence from enemy raids. Helicopters continued to operate even on days when the airfield was bombed and strafed by enemy aircraft.

No regular reports had been received indicating the detailed results of the offensive sorties, due to lack of intelligence and communication facilities then available. Later, from reliable sources it was reported that helicopters operations made invaluable contribution in the operations against infiltrators in particular. On 25th Nov, the Home Minister, Govt of J&K mentioned to the CO of 1 Wing that helicopter operations had caused much more damage and proved far more effective than what had been earlier surmised. He stated that this had been based on the reports which he had lately received through his sources. Staff of 920 Hospital at Srinagar, army commanders of particularly 19 Div and other army units praised the role played by the helicopters. Prompt evacuation of casualties from forward and remote areas was instrumental in saving many lives, besides raising the morale of the troops and giving them the confidence that the Air Force will be able to evacuate them with care and speed. Lying casualties airlifted from forward area were landed directly at the improvised helipad next to the hospital at Srinagar. When it was not considered safe to bring transport aircraft to Srinagar, casualties were flown out from Srinagar and Rajauri by helicopters to the hospital at Udhampur from where they were further evacuated to hospitals in the rear.

Though cease fire was declared on 22nd Sep 1965, helicopter operations from 1 Wing continued with the same tempo. There were infiltrators still to be dealt with. With an uneasy cease fire and many cease fire violations, the logistic support by helicopters in this area was all the more critical. During this period of little over three months that the helicopters operated, they did 1,464 sorties and flew 1,085.30 hours. A brief outline of the effort put in under different roles is given below:

- **Casualty Clearance.** 282 sorties were done to clear 1,323 casualties and 191 hours were flown towards this task. These were mainly from 19 and 25 Div areas. They were airlifted from various helipads/ clearances from the conflict zone.
- **Offensive Sorties.** 80 sorties were done to drop 1,227 bombs (20 lbs) and fire 6,848 rounds on targets given by HQ Sri Force, 19 Div and 25 Div. 86 hrs were flown to meet these demands.
- **Logistic Support.** 935 sorties totalling 688 hours were flown to meet the commitments of different army unit/formations. This includes air dropping and landing of supplies, conveyance of passengers and recce missions. 1,09,637 kg of supplies were air dropped and 1,35,401 kg of landed at forward dropping zones and helipads. 500 passengers were conveyed to various destinations within J&K area. These include Senior Army, Air Force/State and Central Govt officials.
- **Dispersal of Aircraft.** Throughout the operations particularly

during the Indo-Pak hostilities, helicopter flying was so staggered that during the day light hours, the least number of aircraft were at the airfield and at any one time. The dispersal plan of the detachment was executed at the airfield as well as at the helipad area in Srinagar Cantonment. In fact, a small detachment had been organised for the operations of helicopters form this satellite helipad.

- **Accidents.** During the period there were four accidents rendering three aircraft Category E and one aircraft Category B. The Cat B aircraft was promptly patch repaired and flown out from the forward helipad where it had forced landed. The other three aircraft were commendably salvaged to the fullest extent possible.
- **Damage due to Enemy Action.** During the strafing and bombing raids by Pakistan, five Mi-4s were damaged due to shrapnel and blast. Four of these had minor damages which were promptly attended to and aircraft made airworthy. One aircraft was badly damaged, normally beyond the scope of unit repairs. However, it was commendably patch repaired by the maintenance staff in a record time of two days and made airworthy.

During this period, helicopters based at 1 Wing did remarkable work, against heavy odds, silently and for obvious reasons without any outside recognition or publicity. The credit for successful helicopter operations goes entirely to the aircrew and the ground crew who-carried out the task undaunted and undeterred.

1965: Improvisation and Innovation
(A Personal Account of (then) Flt Lt BS Bakshi)

The hurried improvisation of armed helicopters using Mi-4s during August 1965, when infiltrators were detected in Srinagar Valley, is a story unknown to many. At that time I was posted to a unit located in Bareilly and had gone to Bombay to collect two MI-4, being readied by the Erection Unit. Our reaching Bombay and the threat in the Valley coincided. We were of course unaware of the situation, but on 12 Aug 1965, verbal instructions were received to collect the machines and fly to Jammu. The urgency could be noticed from the speeded up assembly of Mi-4; working on a Sunday and at night. Even the Russian specialists were pitching in with all their efforts. We were able to fly out on 15 August 1965 and reach Palam the next day; and, after briefing by Air-II, reached Jammu the same day. By this time, Jammu had become a congregation for Mi-4s from three units. The idea was to use these machines in search and attack missions.

The quick-response good old LMG with 303 Calibre was tried, firing from side windows and from the door removed for the purpose. The barrel was found to be cumbersome, restricting the arc of fire and ineffective when flying beyond the range of small arms fire. Next in series was the 7.62 machine gun, mounted on a tripod position in the door. This was a new weapon at that time as they were replacing the ancient 303 rifles, and soldiers had pinned great hopes on this self loading rifle. During trials, this weapon too had insufficient range and a limited arc of fire besides requiring a mounting spot for tripod. While the trials of field ideas were going on, IAF's technical brains were at work in a BRD. They were able to retrieve the .5 Browning of a Liberator and fix it in the gondola of MI-4. The metal skin of the gondola was replaced by perspex and a slit given for movement of the gun barrel. The gunner could sit astride and fire at targets straight ahead till helicopter was almost over the target. The pilot could visually see tracers and suggest minor corrections. This was found to be a effective weapon. In one case a hut containing explosives was attacked (without knowing of course). The hut blew up on the third burst.

In all these trials, Flt Lt Narayanan, a test pilot, had taken an active part. He had flown Mi-4s during initial induction and had experienced modification to drop 25 lb anti-personnel bombs. Two methods were finalized. Firstly, an attachment under the belly allowing bombs to hang out on outboard stations, which could be released by a switch in the cockpit. While accuracy was good, only eight bombs could be dropped in one sortie. In the second method, a chute was fixed with three vertical channels, each with three bombs. On signal from the pilot, the gunner would release the stop and all bombs would go out. The direction of chute was such that bombs sliding out would be facing about 20 degree towards the tail from the vertical axis. The direction of chute was kept this way to avoid the slipstream coming into cargo cabin. The bombs would tumble on exit and then stabilise to fall vertically down. The number of bombs carried in cargo cabin could be up to a payload of 800kg to 1000kg with crew manually recharging the chute. Moreover, all nine bombs bursting together gave a wider spread of shrapnel. This mode of attack was found to be very effective against column of mules etc. especially when caught on a plateau.

Trials were going on at different places and by 18 Aug 1965 we had helicopter and crews from three different units collected at Srinagar. The task was employing armed helicopters to counter the threat posed by infiltrators. The manpower from different units choked the administration facilities but team spirit carried the day. A total of *567 sorties logging 433 Hrs of flying* were flown from 18-22 Aug 65. Three helicopters crashed, one of them due to ground fire, however there was no loss of life.

Unit Diary Records

109 HU In the month of March 1965, the unit commenced operations from ALGs in Gujarat. Amidst difficult operating conditions for man and machine, border reconnaissance and casualty evacuation was carried out in the Rann of Kutch area. In Jul 65, massive air drop of food and supplies was carried out for marooned troops in the same area. On 15th Aug, a detachment of six helicopters was positioned at Jammu for operations, wherein they flew with distinction in the Pathankot and Chhamb-Jaurian sector. On 21st, this detachment was moved from Jammu to Srinagar for the duration of the war where they were modified for bombing operations. During the operations the unit flew a total of 225 sorties in varied roles like bombing, strafing, casualty evacuation, supply dropping, reconnaissance and communication. One Zero Nine had made its first kill and established itself as a force to be reckoned with.

111 HU The Unit's history is dotted with stories of valour and tales of courage under fire during the war of 1965. It took part in 'Op Orchid' for the large scale troop induction at the Burmese Border. In Mar 1965, it started providing transport support for the Indian Military Training team to all the helipads in Bhutan. The first opportunity to test its battle claws came in Aug 1965. As relations with Pakistan rapidly deteriorated, on 20th Aug 1965, it received orders to move to Srinagar in preparation for an anticipated PAK mischief in that area. By dusk, 10 Mi-4 helicopters and all personnel were ready to move. A Fairchild Packet reached Hashimara that rainy August day and was loaded to capacity with all equipment. On 21st Aug 1965, the 10 helicopters flew in formation under the dynamic leadership of the officiating CO, Flt Lt JL Dweltz. The Air Force decided that all helicopters were to undergo modification of rocket carriage. Technicians worked the whole night to have all the helicopters ready for ferry on the morning of 22nd.

When the helicopters reached Chandigarh, instructions were received to send five helicopters to Srinagar and keep five helicopters back at Chandigarh. Sqn Ldr Kalra, the CO joined the Unit at Chandigarh on 23rd Aug. The inevitable hostilities began a few days later and the Unit helicopters were ready to meet the challenges. From Chandigarh and Srinagar, the indomitable MI-4s launched rocket attacks deep into enemy territory. Two helicopters launching an attack from Jammu came under heavy ground fire while crossing into Pakistan territory. One of the helicopters suffered multiple hits. Not to be daunted by such adversity, the mission was completed and the aircraft returned to Chandigarh. Flt Lt Dweltz and Flt Lt

Dutta piloted the helicopter that was hit while the other was piloted by Flt Lt Goswami and Flt Lt Kanwar. The Unit's five helicopters from Srinagar moved to Jammu on 26th Nov 1965 while the other five helicopters continued at Chandigarh. Four officers of the Unit received gallantry awards and the Mi 4 had drawn first blood in battle.

114 HU Prior to 1965 Operations, 114 HU carried out extensive reconnaissance to detect infiltrators in the Srinagar sector and observation sorties were flown for the Indian Army in Sialkot and Uri-Poonch sectors. Within one year of its formation, the unit was called upon to prove its mettle in the 1965 operations. In the build up phase of the 1965 war, numerous sorties were flown with various Army Commanders including the then Chief of Army Staff. During the operations, the unit helicopters were a familiar sight, with its Chetaks festooned with large Red Cross markings, undertaking casualty evacuation and mercy missions. In acts of real heroism, 114 HU pilots evacuated 46 casualties from the battlefield in the midst of intense enemy fire and frequent air raids between 11 and 23 September 1965. In all, the unit flew 105 hours towards evacuation of 79 battlefield casualties and towards intelligence gathering missions.

A Tribute to Indomitable Courage & Innovation: Gallantry Award Citations

Sqn Ldr Bhagat Singh Bakshi (4597) General Duties (Pilot)

Squadron Leader Bhagat Singh Bakshi took over as officer in charge of a helicopter detachment drawn from three units. The detachment carried out 567 sorties and flew 433 hours within a period of thirty six days from 18th August to 22nd September, 1965. The aircraft operated immediately behind the lines carrying out supply-dropping, bombing and strafing. They evacuated eight hundred and thirty two casualties. They dropped a large number of bombs and fired many rounds of ammunition and conveyed large quantities of ammunition and supplies. This was of vital help to our advancing forces to sustain the newly captured pickets in Jammu and Kashmir. On 31st August, 1965 when he was on an offensive support mission, he had to crash-land his aircraft because of a technical failure at a height of about 10,000 feet in an area infested by Pakistani infiltrators. The crew escaped with only minor injuries, thanks to his flying skills. He walked 18 miles helping the injured crew and organized the evacuation of those who could not walk. He carried out three offensive sorties against the infiltrators, flying at low levels while attacking the targets. He also carried out thirty logistics sorties and thirteen sorties to evacuate casualties from quickly fabricated helipads within the range of enemy fire. These sorties were flown from an airfield which was under the constant threat of enemy air raids without any warning. Undeterred by such grave dangers, the officer accomplished his difficult task with great determination and zeal and displayed a high degree of professional skill and ability which were in the best traditions of the Air Force.

Flight Lieutenant John Leo Dweltz (5707), General Duties (Pilot)

During the operations against Pakistan, Flight Lieutenant John Leo Dweltz, Flight Commander of a Helicopter Unit, flew twenty offensive, thirty six logistic, twelve casualty evacuation and four reconnaissance sorties within a period of thirty six days. He operated from quickly improvised helipads in difficult terrain with utter disregard for his personal safety. By evacuating casualties from forward posts he saved many valuable lives. His supply-dropping missions helped to maintain the newly occupied picquets. He continued to operate from helipads within the enemy's shelling range and under constant threat of enemy air raids on his airfield without warning. On 11th September, 1965, Flight Lieutenant Dweltz was detailed to attack a strong hold of infiltrators at Raman Nallah. While he was bombing and strafing the enemy bunkers in a narrow valley, the infiltrators started firing at his aircraft with small arms. Undeterred by the enemy fire, he carried out his mission with determination. On 15th September 1965, the army reported strong enemy pressure on some of our posts in the Tithwal area and wanted the Air Force to destroy a rope bridge over the Kishan Ganga. The bridge ran through a narrow valley and the area was occupied by Pakistani infiltrators. He made three runs over the bridge and dropped forty eight bombs on the target. He set fire to both ends of the bridge and caused considerable damage to its centre portion, thus severing the line of communication of the enemy and relieving pressure on our picquets.

Flight Lieutenant Subhash Madanmohan Hundiwala (6351) General Duties (Pilot)

Flight Lieutenant Subhash Madanmohan Hundiwala has been in a helicopter unit in Ladkah since Novemeber, 1963. Since then he has flown 430 hours on operational sorties in that area. In spite of the hazards of flying over difficult and mountainous terrain of Ladakh, he has always volunteered for difficult operational missions allotted to this unit. On 18th May, 1965, he undertook four sorties in adverse weather conditions to evacuate army battle casualties from Kargil. On 28th May, 1965, while on a sortie to evacuate a serious casualty from a place in the Karakoram ranges, he experienced a flame-out of the engine; yet by skilful handling of the helicopter, he carried out a successful forced landing on a helipad situated at an altitude of 15,000 feet. Throughout his tenure of duties in Ladakh, this officer has successfully carried out difficult missions assigned to him and has displayed courage, professional skill and devotion to duty.

Flight Lieutenant Lalit Kumar Dutta (6506) General Duties (Pilot)

During the operations against Pakistan, Flight Lieutenant Lalit Kumar Dutta, a senior pilot of a Helicopter Unit, carried out eighty one sorties and flew fifty seven hours under very adverse conditions and from quickly improvised helipads, just behind our advancing troops and within the range of enemy's small arms fire. At great personal risk, he carried out eighteen offensive sorties, bombing and strafing the strongholds of infiltrators effectively. He flew twenty five sorties evacuating sick and wounded soldiers, thus saving many valuable lives. He also flew thirty sorties to convey ammunition and essential supplies for sustaining forward pickets. All this while, he operated from an airfield which was under constant threat of enemy air raids without warning.

Flight Lieutenat Ranjit Kumar Malhotra (6513) General Duties (Pilot)

Flight Lieutenat Ranjit Kumar Malhotra has been flying Allouette III helicopters in Ladakh since Jun 1964 and has flown about 400 hours in Ladakh and other areas of Jammu and Kashmir. Despite difficult and mountainous terrain he has always volunteered for difficult operational missions and set a fine example to his colleagues. In May and Jun 1965, in adverse weather conditions, he undertook twelve sorties to Kargil for the purpose of evacuating our army battle casualties. Again on 14th August 1965, he undertook two sorties to evacuate army battle casualties from Naugam in Srinagar valley. During these sorties, he had to land at a place which was encircled by Pakistani infiltrators. They fired at him after he had landed but with cool courage and presence of mind, he successfully evacuated the casualties from the encircled area. The courage and devotion to duty displayed by him are in the best traditions of the Indian Air Force.

Flight Lieutenant Chuhar Singh Kanwar (6532) General Duties (Pilot)

During the operations against Pakistan, Flight Lieutenant Chuhar Singh Kanwar, a senior pilot of a helicopter unit, carried out fifteen offensive sorties against the infiltrators, which were very effective and of immense help to the troops engaged in the mopping up operation. He carried out forty five logistic sorties and twenty casualty evacuation sorties in a period of thirty six days. He operated from quickly fabricated helipads, right behind the line of actual control and carried out four trial landings on these helipads. By evacuating casualties from the most forward areas, he saved many valuable lives. His supply missions were a life line for troops fighting in difficult terrain. He operated from an airfield which was under constant threat of enemy air raids without warning. On 13th September 1965, during an air raid, a helicopter was badly damaged. The engineering staff carried out some repairs; but there were still a number of navigational and functional limitations. He volunteered to airtest the aircraft and finally ferried it to a safer airfield and saved it from further damage by enemy action. On 22nd September 1965, when he landed at an improvised helipad with vital ammunition and essential supplies, the enemy started shelling the helipad. The officer unloaded the supplies undaunted by enemy shelling. With courage and presence of mind, he then took off quickly and brought the aircraft and aircrew safely to the base.

Flight Lieutenat Premananda Goswami (Aux. 30082) General Duties (Pilot)

During the operations against Pakistan, Flight Lieutenant Premnanda Goswami, flight commander of a helicopter unit, flew seventy sorties in a period of thirty six days and completed sixty one hours of flying. He flew twelve casualty evacuation sorties from quickly improvised helipads, just behind our advancing troops and within the range of enemy shells. The hilly terrain was very difficult and treacherous for helicopter operations. Undeterred by these difficulties and dangers, he carried out his mercy missions and saved many valuable lives. He also undertook sixty logistic sorties supplying and drooping ammunition and rations to forward posts at great personal risk. He operated from an airfield which was under constant threat of enemy air-raids without warning. On 22nd September 1965, he landed at an improvised helipad with ammunition and supplies to sustain this picket. The enemy started shelling the post. With courage and presence of mind, he took off while the other aircraft was still unloading. Disregarding his own safety he circled near the enemy positions to divert the fire and attention of the enemy to this own aircraft, in order to give time to the other aircraft to unload its cargo and save it from destruction.

CHAPTER V

The Birth of Bangladesh – 1971

Introduction If there is any home-grown campaign in India that truly boosts the concept of jointness, it is the war for liberating Bangladesh. The multiplying effect of synergy between the three services created magic on the eastern front while containing damage in the west. The whole gamut of joint support structures such as Advance HQs, Tactical Air Centres (TAC), Forward Air Controllers (FACs), Ground Liaison Officers (GLOs) and others worked in a coherent manner to achieve air superiority in the east and effective offensive air support (OAS) in the west. The Blitzkrieg on the ground in East Pakistan ably supported by a well coordinated air campaign demonstrated the Air-Land Battle concept well before it came into common parlance in the Eighties and Nineties.

Wars, it is said, are fought in the minds of the generals, air marshals and admirals. Nothing could be truer than what was demonstrated in the lightening moves of the Indian Army/IAF combine in the east. Learning from the Vietnam campaign, the IAF top brass in collusion with Lt Gen Sagat Singh, GOC 4 Corps, conceived a series of Special Heliborne Operations that leap frogged the Indian infantry to the centre of gravity of Gen Niazi's forces. The early surrender was due to the speed of advance that had even taken the Indian Army HQ by surprise! This was surely the best example what mutual respect and faith in each other's capabilities could bring about among the Indian Armed Forces.

While the main focus of this chapter is on the pioneering SHBO missions that redefined tactics, commanders at all levels displayed remarkable ingenuity and creativity to grab every opportunity for denting the enemy's might – special armed operations by helicopters along with the Mukti Bahini is one such story. Though a drop in the ocean in the overall scheme of things, the effect it had on the morale of local fighters and populace was immeasurable. Recent campaigns of NATO list local support as a lynchpin for success; Indian armed forces demonstrated it well in 1971. The successful large scale airborne assault at Tangail was the icing on the cake of what true jointness could bring about. Rarely in history, has a more coordinated and successful airborne assault been recorded than done by the IAF-Indian Army combine in 1971. The gallantry citations at the end of the chapter only serve to flag the commitment of the IAF helicopter fleet to the joint campaign in general, and the Indian soldier in particular.

The Air Force Strategy

The Indian Air Force entered the war with Pakistan with separate strategies to conduct air operations in the western and eastern theatres. While the strategy for the western theatre was directed more towards a holding action. An aggressive strategy was formulated for the eastern theatre. This strategy dictated that:

- Military operations would be mounted with the specific aim of an early liberation of East Pakistan.
- To annihilate the Pakistan Air Force and attain total air superiority in this theatre.
- Launch unabated offensive air support (OAS) missions in support of the Indian Army.
- To assist the Indian Navy in isolating East Bengal and to ensure that the PAF did not hinder our naval operations.

With the ground situation stabilised in the Chamb sector from 14th Dec 1971, WAC concentrated on denying the Pak Army all forms of movement in the rear of their troops facing 1 and 11 Corps. Offensive sweeps took on anything that moved on road or rail. A section of the railway line near Khudian was completely destroyed by S-22s. At the same time, close support to Indian troops continued in all sectors. One army Air Op Krishak aircraft was shot down by an F-86 near Nainakot. An IAF Mi-4 helicopter already airborne from Samba was diverted to the crash site, and the two casualties were flown to Udhampur within an hour

Action in the East

HQ, Eastern Air Command was located at Shillong and the Advance HQ, EAC was co-located with the Eastern Army Command in Calcutta. The AOC, Advance HQ was responsible for the conduct of all offensive air support (OAS) operations and limited Transport/Helicopter operations in areas west of the River Meghna and all operations east of the river were controlled directly by HQ 4 Corps and HQ 33 Corps. HQ 101 Comn Zone located at Tura was supported by an ad-hoc TAC at HQ, EAC. The effective strength of IAF combat aircraft as on 3rd Dec was 623 aircraft which did not include transport and helicopter assets. HQ EAC was allotted the following force level between 3rd and 17th Dec: 143 AD & Strike aircraft; 12 Bombers; 78 Transport aircraft; 82 Helicopters.

The type-wise breakdown of transport aircraft and helicopters under the command of HQ, EAC for conduct of air operations in the Eastern theatre in Dec 1971 was: 32 Dakotas; 18 Caribous; 12 Packets; 16 Otters; 80 Mi-4s; and, 22 Allouettes.

Summary of Flying Effort: 3rd–17th Dec 1971

Type of aircraft	Sorties	Hours	Load	Passenger carried	Casualties carried
Mi-4	1397	992:05	111.1	4065	866
Allouette	894	726.35		809	282
Bell	113	84:00			31
Total	2404	1802:40	111.1	4874	1179

The Battle for Bangladesh

Concept of Ops: Indian Army The Army Commander was aware that the battle in East Pakistan would be a fluid one and that his operational instructions should be used as a basis for general guidance. Though final objectives would remain unaltered, formation commanders were to be prepared to constantly review thrust lines and intermediate objectives as per the developing situation. These modifications involved by-passing sizeable enemy strongholds, which not only indicated a significant change in concept, but also expressed a growing confidence and realization by Eastern Command that a swift advance would threaten Dacca and bring about a total collapse and defeat of Pakistani forces. Initial plans catered for securing the line of the Brahmaputra and Meghna rivers. It was planned that 4 Corps would cross the Meghna between Daudkhandi and Bhairab Bazar and advance to Dhaka. 101 Communication Zone Area with 95 Mtn Bde Gp of four battalions, 2 Para Group, followed by 167 Mtn Bde, would advance to Dhaka from the North. These forces would be supported in their advance to Dhaka by Siddiqi Force, a large group of freedom fighters which had been operating with good effect in Tangail district.

Contingency plans were also made for elements of 2 Corps to cross the Padma at Goalundo for which purpose suitable inland water transport craft were placed at Dhubri to assist 95 Mtn Bde or move elements of 20 Mtn Div, if required. Army Headquarters reserve of 50 Independent Para Bde was to provide one complete battalion drop and two company drops. It was appreciated that the most important area for the main drop was Tangail in order to ensure the early capture of Dhaka. Second priority was given

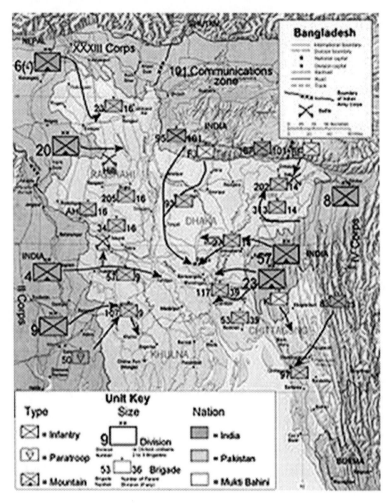

for two-company drops to assist in securing Magura if necessary. Due to
the limited availability of Mi-4 helicopters, all these helicopters were
allocated to 4 Corps to enable them to ferry troops as required. The air effort
allotted to the Army was sufficient to cater for an average of 120 sorties per
day. Naval forces were to blockade Chalna and Chittagong and Naval Air
was to be used in the area Chittagong-Cox's Bazar.

Mi-4s Mi-4 units had seen continuous operational service in India's
northeast, supporting the Indian Army's counter-insurgency operations in
Nagaland and Mizoram; lifting commando teams into action and pre-
empting hostile infiltration to and from China. They had often suffered hits
from small arms fire; on one occasion, a Mi-4 even returned to base with a
poison-tipped arrow sticking to its undercarriage! The operational climax

for the Mi-4s in IAF service came during the 1971 war when this helicopter played a key role in the Indian Army's Blitzkrieg-style advance to Dacca. Hostilities had broken out on 3rd December, and a lightning campaign was unleashed in the eastern sector.

Whilst the Indian 2 Corps battled its way from the western side, the 4 Corps thrust came from three directions in the east, across 250 km of border between the erstwhile East Pakistan and Meghalaya to the north, through to the Feni Salient in the extreme south of Tripura. One column moved in from Silchar-Karimganj area towards Sylhet, another along the Akhura-Asjuganj axis and the third in the south with the objective of containing Comilla and cutting off Chittagong. Whilst battling against the two Pakistani divisions ranged against them, 4 Corps had to contend with the most formidable riverine terrain, criss-crossed by rivers and streams that make the area a logistic nightmare for any offensive action.

Supporting the advance was 110 HU, with 10 Mi-4s on strength and another two Mi-4s attached from 105 HU. The Mi-4s were intimately involved in the spectacular progress made by the Army, particularly 8 Mountain Division. 110 HU transported troops, guns, ammunition and equipment, evacuated casualties and conducted reconnaissance for the field formations in the sector. Between 3rd and 6th December, numerous sorties were done to position for the coming Blitzkrieg. Many daring casevac missions were carried out from the tactical battle area (TBA). 22 SHBO missions were flown on 7th for the Battle of Sylhet. 584 troops and 125 tons of supplies and nine field guns were lifted on 9th and 10th December to reinforce the build-up against Sylhet, which was held by the Pakistani 313 Brigade. 110 HU then moved to Agartala for the heliborne operations from Brahmanbaria to Raipura.

Mi-4s were the focal forces which enabled the pace of the Army's relentless drive that now saw heliborne infantry leap-frogging past natural obstacles and enemy defences; with MiG-21s and Gnats strafing and rocketing strongholds, in a text book display of co-ordinated air power. The biggest operational task yet undertaken by the IAF's helicopter force was between 11th and 15th December. On the 11th, 1,350 troops and 192 tons (195tonnes) of equipment were airlifted from Brahmanbaria to Narsingdi, including in particular the lead battalion 4th/5th Gorkha Rifles (FF) of 59th Mountain Brigade, spearheading 8th Mountain Division's intrusion in Sylhet. The Gorkhas and their equipment were lifted across the Meghna River to south-east of Sylhet, forcing the much larger enemy force to abandon defences and, for fear of being cut off by the heliborne forces, pull back into the town. On 14th December, the 12 Mi-4s were positioned at Dandkandi to airlift, across the Meghna to Baidya Bazar, 810 troops and 23 tons of equipment; and in another major effort the next day, they airlifted 1,209 troops and 38 tons of equipment to strengthen the Baidya Bazar complex. A total of 4,500 troops and 515 tons of equipment were helilifted by the 12 helicopters in this action.

Heliborne Operations In the Eastern sector, from 7th Dec onwards, helicopters began transporting troops, light guns, ammunition and POL. In three days of operations, the helicopters had flown over 120 sorties to lift nearly 1200 troops and 10,000 kg of load. Over 50 sorties were carried out during night. These operations facilitated capture of Sylhet and certain other features in the same sector. The night movement of troops in large numbers caught the Pak Army by surprise. The biggest operations, however, took place later. On 11th there were 161 sorties which carried nearly 1300 troops and 6000 kg of load. The load included RCL guns, gun shells and mortars. The next day there were 1140 sorties which airlifted over 2800 troops and 33000 kg of rations and ammunition. The heliborne operations continued at a rate of over 20 sorties per day up to Dec 17th, 1971. The helicopter force of the IAF could fairly claim good credit for the speed of advance of the Indian Army in Bangladesh.

As has been well recognised by the Indian Army, IAF Mi-4s were the key factor in the lightning campaign, where speed of advance was of the utmost importance. Other squadrons that were given special mention were 15 and 24 Squadrons whose Gnats provided escort to the heliborne operations and close support over the entire Brahman Baria-Sylhet axis. Other Mi-4 and Allouette III units, including 112 HU in the Jessore sector, conducted casualty evacuation, communication and reconnaissance flights. Some Mi-4s and Allouette IIIs were equipped as gunships and operated in this role both in the east and the north-west of India, recalling their earlier role in 1965, when they hunted infiltrators with machine guns and anti-personnel bombs in Kashmir.

SHBO Sylhet On 7th, the order was to airlift one battalion of troops to Sylhet by 1200 hrs. The Mi-4 task force moved to Kailasahar. After the initial recce and sighting of the destination helipad, airlift started by 1500 hrs with the jump off point as Kalaura. By evening, 22 sorties were carried out, airlifting 254 troops and 400 kg of ammunitions in the face of heavy enemy ground fire. Later at night, a special task involving three more sorties were carried out to lift some vital signal equipment. SHBO Sylhet concluded on 9th Dec after 584 troops and 12500 kg of load had been dropped. Nine helicopters suffered damage from enemy ground fire,

SHBO Raipura On the 9th, immediately after SHBO Sylhet concluded in the morning, the task force commenced airlifting troops of 4 Corps from Brahman Baria at 1600 hrs; destination was Raipur. This operation possibly saved the Indian Army 26 days of marching through riverine terrain, bogs and swamps, not to mention the man-made defences.

SHBO Narshingdai Operation Narshingdai commenced on 11th at 0530 hrs. The kick-off base briefed was Brahman Baria and the target DZ agreed was Narshingdai. Operations continued till 13th Dec, and it was the biggest SHBO operation during the 1971 War.

SHBO Baidya Bazar The task given was induction of troops into Baidya Bazar. Operations commenced despite difficulties and delays by over 10 hours and 30 minutes on 13th Dec, and by 15th at cease of operations, 1209 troops and 38100 kg of load had been lifted. Helicopters also undertook casualty evacuation, which involved picking up injured troops straight off the battle front in the face of enemy ground fire.

The Mukti Bahini Air Force The birth of the Bangladesh Air Force as Kilo Flight has hardly been noticed. On the night of December 3rd and 4th, Otter aircraft of this unit attacked fuel dumps at Chittagong and an Allouette helicopter raided the fuel dumps at Narayanganj. Both missions were successful. On 4th December, this flight at Kailashahar airfield was placed at the disposal of GOC 8 Mountain Division operating in the Maulvibazar/ Sylhet sector. The GOC asked for attacks by night on convoys, river barges and streamers on the Meghna, north of Bhairab Bazar. It flew five sorties between 4th and 7th, hitting bunkers and troops concentrations at Maulvibazar, and also destroying two steamers and several 3-ton trucks carrying troops. From 7th to 12th, the Allouette helicopter was constantly and effectively used as an armed escort during the special helicopter operations at Sylhet, Raipur, Narsingdi and Baidya Bazar. It engaged targets as directed by the FAC at night also. After the first night raid on Chittagong fuel dumps, the Otter aircraft was utilized for armed escort/ recce on a few missions only in the same areas the Allouette. The overall returns from this Flight were not very encouraging, perhaps because it had no definite tasks. Contribution of this flight to the war effort was hardly significant. However, as far as raising the morale of the Mukti Bahini was concerned, it did serve the desired purpose. During the period of operations the total flying effort of this flight was: Allouette – 77sorties /68 hrs 10 min and Otter – 12sorties /23 hrs 35 min

First-hand Account of Pushp K Vaid

3rd December 1971: Official war declared against Pakistan. I was posted to

110 Helicopter Unit at Kumbhigram. Our unit along with 111 & 105 HU was located at Agartala Airport. We had 14 helicopters altogether. I was the most senior flight commander, so I was in-charge of running the show. We were moved in the morning to a field about 20 miles from Agartala. The GOC briefed the CO and a couple of other officers on what the plans were for the day and then took us along in a helicopter to show us the field near Sylhet where he wanted us to drop the formation that day. We were ready from 0900 hrs and were waiting for the army to appear. It had been held up enroute so they only started appearing at around 1500 hrs.

Our briefing was to get airborne as soon as the army arrived. I had allocated the crew for each helicopter and each one of us had a number. I was number One and second helicopter was number Two and so on. Idea was for the first helicopter to get airborne, as soon the helicopter was full of soldiers & material etc. followed by the second helicopter and then the third and so on. Sylhet helipad was 20 minutes flying from where we were. The helipad was a big field perhaps 20 acres. I got airborne at 1500 hrs, and the other helicopters got airborne one after the other at an interval of 2 to 3 minutes; so by the time I had dropped my load and come back for the second load the last helicopter was just getting airborne. We had enough fuel to do three trips each. But by the time I was going for my third trip it was already dark. Also, the Pakistan Army had seen the helicopters and started surrounding the field we were landing in and were firing at us as we came in to land. On my third flight I must have seen as many as five hundred bullets or more coming towards me. I couldn't count of course but they looked like a lot of bullets along with tracers. Amazingly not one bullet hit us. I warned all others pilots to keep a look out and to reduce the time on the ground to minimum. We all finished our three trips and shut down for fuel at our bases. All of us checked our helicopters and not even one bullet had hit any helicopter. So much for Pakistani marksmanship!

While we were getting ready to start flying again, Group Captain Chandan Singh, who was in charge of the Air Force element, said that we won't fly at night and start flying next morning. That really upset the army. They had not planned their soldiers with ammunition, food or radios. They were not even sure of what or who had gone so far. All the helicopters had been filled up as the soldiers arrived with the idea that we will keep flying till everything had been moved. There was a big discussion between the Army and the Air Force. Eventually, after nearly three hours, it was decided that one helicopter will go and if it came back the others will be sent.

Gp Capt Chandan Singh asked me to get a crew for the flight. I came to our rest room and asked for a volunteer to join me. Every one of my forty pilots wanted to come. I picked Flying Officer Kanth Reddy. We filled the helicopter with the radios and other equipment along with the soldiers. For our protection we had a .38 revolver and a Lungi. The Lungi was to pretend we were Bengalis in case shot down. We always carried a flight engineer who could do all the handling on the ground, so that we did not stay on the ground too long. So off we went at about 2200 hrs. Pitch dark. Our only navigational aid was a compass. We set course and as we got closer we heard our air force liaison officer, who had gone along with the army, on his hand held radio. He would not tell us what the situation was on the ground in case we turned back. He told us that it was all quiet and that he would light a fire for us so that we knew where to land. As the fire was lit, the entire Pakistan army started firing at that fire. We could see tracer bullets, hundreds of them. The poor soldiers who lit the fire got hit.

We landed and dumped everything we had as soon as we could, picked up the soldier who had been hit and got airborne in a few seconds. After all the excitement of bullets, I asked my flight engineer to have a look at the helicopter and see if we had been hit – not one bullet – amazing! On my way back, I radioed the base to get all the helicopters ready so that we could continue flying right through the night. I landed back, picked up the next load and went again for the drop. We then came back and picked the third load before the rest of the helicopters got ready. On way back from our third trip, we noticed that our fuel gauge was showing zero fuel. We knew we had taken off with enough fuel, so either we had a leak or the gauge was not working. We had fingers crossed; legs crossed and were praying to whoever is up there to get us back to base. Well we got back to base, thanks to the LORD. At base we discovered that the cable to the fuel gauge had been hit. This was mended and then we continued flying with the rest of the helicopters right through the night till we had dropped the entire brigade at Sylhet.

First-hand Account of Air Cmde CM Singla

During 1971, I was a Flight Lieutenant serving as one of the instructors in type conversion unit for Chetak helicopters. War was imminent. I was asked by my CO to proceed to Tezpur. Task was not specified. Not wanting a desk job and separation from my flying unit at the critical juncture, I protested – but was overruled. At Tezpur, the station Commander called me over. I was asked to take one helicopter from the station and fly it alone to Dimapur. I was advised that details would be provided there. Full maintenance support was assured on top priority. At Dimapur, the AOC Jorhat arrived in a Dakota. Some foreigners were already located at Dimapur. I was introduced to them by the AOC. They were transport, bomber and fighter pilots. I was asked to convert them to Chetak helicopter by day and night. Deadline for making them fully operational was set. I was directed to forget IAF rules and syllabi etc.

We were given a cot each in the ramshackle flying control and were advised to eat with the men. The base then had only one living – out officer and about a dozen airmen. My pupils had an opulent background and resented the poor quality food and lack of basic creature comforts. Further, they were under severe mental stress. Add to this their fixed wing background and age factor, and the potential environment was not rosy. However, we flew rigorously and deadlines for day and night conversions were met. Soon after, I was asked to pick up an armed Chetak from Tezpur which had just been modified by and airlifted from BRD. It had one rocket pod on either side and each pod carried seven rockets. These could be fired in pairs or in salvo. In addition, on the helicopter floor a twin barrel machine gun was mounted.

We were now required to learn to fire accurately by day and night. Practising on the machine gun was not a problem, but for the rockets we needed a gun-sight, harmonization and above all, targets. I decided that bundles of hay would be para dropped over the thick jungles. The parachutes would form a canopy on tree-tops and these could be used as targets. However, the parachutes just went through the foliage. Ultimately, I hover dropped men on grassy hill slopes who hammered parachutes on the ground. These presented fair targets. We burnt several parachutes and became fairly proficient at firing rockets by day and night. Our first operational task came soon after. We moved to the vicinity of Kailashar and launched the armed Chetak by night from the wilderness. To overcome recovery problem on a pitch black night, judicious use of concealed Petromax was made. Enemy fuel dumps were successfully set afire by rockets that night.

Then onwards, we operated as a self contained and mobile hitting outfit. We moved frequently to support troops as they advanced. I was ordered to fly in command on all armed missions. On one night, troops were to be inducted by Mi-8 helicopters. I was to give air cover. The Forward Air Controller told me on R/T that he was my pupil at Bidar. Our effective rocket firing over several missions that night was witnessed by a senior officer who flew as a passenger with us initially. Being an unstable platform, in our dive to aim and engage targets with rockets pairs, we had to come pretty low and close to the targets. Thus we stayed for long durations over hostile territory at low speeds and height. We took some bullet hits that night. Subsequently, we hit & burnt ammunition trucks, boats ferrying troops by night and enemy concentrations. The helicopter had by now got pierced by bullets all over. The engineers did a good job of repairs and despite vibrations etc, we remained operational throughout the war. On one occasion, I was sent to try and retrieve a downed fighter pilot, though he could not be located. After signing of the ceasefire, I was asked to airlift my students to their Capital and leave the armed helicopter with them as a gift. The grateful students placed a case of Scotch whisky in my unarmed helicopter – which of course did not see light of the next day.

Personal Account of Flight Commander 109 HU

I was Flight Commander of 109 Helicopter Unit located at Jammu (J&K) and had eight helicopters (Mi-4) to undertake multifarious assigned tasks. By 1971, I had completed four years at 109 HU and knew the operational flying area very well, which came handy during recce sorties during the forthcoming war. During the period of imminent hostilities when the air was heavy with anticipation, skirmishes were becoming a regular affair. 109 HU had been assigned the important commitment of enabling the Field Commanders in the general area of Pathankot, Jammu, Rajouri and Poonch to undertake heliborne recce from Pathankot to Poonch all along the IB & CFL. The task of flying the commanders (Div Cdrs, Bdr Cdrs, CO's and even company commanders) to as close as 500 m from the border was indeed immensely demanding; with delineation on ground of the international border making the aeronautical maps unworthy for reference. The pillars on the border were the only guidelines. At times when we thought we were well within our airspace, the slant perspective of the Pak troops made them believe otherwise. What would follow was small arms fire, which we learnt to regard as of nuisance value.

After completion of numerous recce flights, one day I was called by AOC J&K, who enquired if any changes were required for undertaking recce in view of the enemy fire. I told him, no change in profile is required as no helicopter had suffered any damage and we continued with mission after mission without any incident. Early October 1971, 109 HU was moved to Udhampur. For routine servicing the helicopters would fly back to Jammu. Before the hostilities, I had flown 38 border recce sorties with army field commanders on board. Never ever prior to hostilities or during hostilities was any armament/ammunition carried on Mi-4s though provisions existed. During one such recce the commander on board expressed keen desire to identify the exact nature of activity reported in the vicinity of sprawling grove in Chamb Sector. The concerned area appeared few kilometres away and instinctively I manoeuvred the helicopter towards the grove. Spurred by my curiosity, I pressed on and arrived over the dense grove. It was a venue for an important meeting of the enemy's field commanders, with a number of jeeps, camouflage nets, guns and troops all over the place. While we were surprised, we did want to make intelligence capital out of the inadvertent straying, and therefore, carried out a low circuit. This outrageous act inflamed and panicked the gathering and soon we saw an intense flurry of activity with jeeps pelting in all directions. After having a good look, I turned towards Chamb and opened full power. As expected, the ground fire opened up from all the six posts on the CFL. A post landing inspection revealed no bullet hit, but it was a providential escape.

There was only one helicopter at Udhampur while the rest were dispersed. I was flying the Corps Cdr almost daily to Chamb / Jaurian sector. On 10th Dec 71, the 15 Corps Cdr, Lt Gen Sartaj Singh was to visit the theatre in the Chamb sector. We took off from Bikrampark helipad with air raid warning on. On our way to Jaurian, I sighted in the distance, nine Pakistani fighters. Being most vulnerable at this stage, I hugged the ground and decided to press on. The Corps Cdr's face registered surprise constantly, while trying to remain as unobtrusive as possible, I continued. Soon we were relieved when we realized that our helicopter – a sluggish prey in these circumstances had escaped attention. The Pak fighters were on their way to attack Akhnoor Bridge. As we reached Jaurian, the magnitude and ferocity of the battle raging in the entire sector was revealed. This was the day when 20 Pak tanks had crossed Manawar Tawi towards Jaurian. The incessant air and artillery fire was nerve-wrecking.

On my return flight from Jaurian, 15 very serious casualties were evacuated to Udhampur military hospital; a mission which was carried out uneventfully. The following day, on 11th Dec 71, another request followed for evacuation of eight war casualties. This was a demanding task as the casualties had to be airlifted from a quickly improvised helipad at Nainkot, the captured territory in Shakargarh sector. Nevertheless, in spite of very difficult navigation, we landed there only to learn that the seriously wounded were dispatched by road. Sensing the arduous and painful journey the injured must be braving to the hospital, we took off to intercept the ambulance en-route. Having spotted the ambulance, we looked for a favourable landing ground. The mine-infested areas sprawled on both sides of the kutcha road – remember it was occupied by the enemy earlier. With extreme precaution, we landed on a narrow track, resting on the tyre marks created by the infantry vehicles. The casualties were transferred, airlifted and safely conveyed to the Pathankot military hospital. It was heart-warming to learn that not a single wounded succumbed to injuries of the battle. After the war, I found a Vir Chakra being added to my small tally of medals.

Unit Records

Desert Hawks: 107 HU In Oct-Nov 1971, 107 HU flew about 320 hours in the preparatory phase. Between 3rd and 17th, it flew about 140 missions airlifting more than 250 passengers and 19000 Kg of load to forward areas amidst heavy firing. A number of POWs were also airlifted by unit helicopters to Jodhpur for interrogation. December 1971 was a memorable month. As the army advanced, vital supplies like arms & ammunitions, food, water and fuel were delivered and a large number of casualties were evacuated. It evacuated more than 100 casualties and airlifted more than 250 passengers, besides airlifting 19000 kg of load to forward areas amidst fierce fighting. On 11th, an unforgettable incident occurred when a helicopter of the unit was on a food-dropping mission deep inside Pakistan in support of Madras Regiment troops. A pair of F-86 Sabres flew directly above the helicopter and soon turned to attack. The enemy fighters kept firing barely missing the helicopter. Soon two more enemy fighters joined the attack. The pilot decided to force land and as soon as the helicopter landed, a bomb fell just missing the rotor disc. The helicopter returned to Uttarlai with its windscreen and window shattered due to an impact close to it but otherwise undamaged. This was perhaps the first of the tactics employed by helicopters for aerial combat against a fighter in the IAF!

Pioneering SHBO Missions in War

Vanguards: 110 HU 110 HU was raised on 19th Feb 1962 with Mi-4 helicopters at Tezpur. During the 1971 war for liberation of Bangladesh, 110 HU was utilized for 'Vertical Envelopment' extensively. It planned and executed the first ever SHBO mission in war. The famous Heli-Bridging operations from Sylhet, Raipura, Narsingdi and Baidya Bazar in erstwhile East Pakistan, shaped the battlefield and endorsed the flexibility and mobility of air power in battle. The fast mobilisation enabled the liberating troops to reach Dacca within 12 days. Thus, the Vanguards emerged as the pioneers of what was soon to become the primary role of the medium lift helicopters of the Indian Air Force-SHBO.

SPECIAL HELI BORNE OPERATIONS AT EAST PAKISTAN

The area of responsibility along the eastern theatre of operations was entrusted to HQ 4 Corps and stretched along the entire length of eastern International Border of East Pakistan. In the course of deliberating the offensive of 4 Corps, the GOC, Lt Gen Sagat Singh decided to launch the Corps offensive in East Pakistan from Northern Tripura and Agartala. Helicopters of 110 HU were initially tasked by HQ EAC for casualty evacuation and air logistics along the axis of thrust. Thereafter, it changed the face of the TBA for the Indian Army and flew the first ever Heliborne Operation during the 'Battle of Sylhet'. This resulted in 4/5 Gorkha Rifles (Frontier Force) mobilizing at a lightning pace, eventually leading to the fall of the Pakistan Garrison at Sylhet. As the battle progressed, the troop mobility east of Dacca was totally reliant on heli-bridging for obstacle crossing over river Meghna and this is where 110 HU contributed by carrying out critical SHBO missions.

Sylhet town was a district and communication centre, and an important place from the military point of view as its fall would have inflicted a severe setback to Pakistan with subsequent international repercussions. 4 Corps tasked 4/5 Gorkha Rifles of 59 Mountain Brigade to capture Sylhet and inducted Mi-4 helicopters of 110 HU to undertake SHBO missions to accomplish the ground plan. The main task was to heli-lift a battalion (4/5 GR) to Mirpara (outskirts of Sylhet) to secure the railway bridge over Surma River and occupy Sylhet airport/radio station to further operations. The challenges for the Unit helicopters did not end with Sylhet and the Unit was subsequently tasked to undertake extensive SHBO missions for furtherance of Indian Army objectives as the battle progressed from different axes at Raipura, Narsingdi and Baidya Bazar. The following extracts from the unit diary details the major missions flown during the 1971 Operations:

3rd to 6th Dec 1971 Three helicopters moved to Teliamura, while one was inducted at Aizwal and two at Imphal for communication and air logistics roles. Two more helicopters were subsequently inducted at Teliamura resulting in a total of five helicopters operating there. All the helicopters were utilized for aerial reconnaissance and casualty evacuation and 57 casualties were evacuated on 4th Dec from the battlefield. These helicopters

THE TOP BRASS INDIAN ARMY LANDS IN A MI-4 OF 110 HU AT EAST PAKISTAN

evacuated 42 casualties from the tactical battle area under enemy fire on 5th Dec and 36 casualties on 6th Dec.

7th Dec – Battle of Sylhet Five helicopters moved from Teliamura to Kailashahar for airlifting a battalion of troops to a point north of Surma River virtually in full view of Sylhet. Three more helicopters which were operating from Aizwal and Kumbhirgram were moved to Kailashahar. The helicopters flew 22 SHBO missions in three shuttles and airlifted 254 troops and 400 kg of ammunition. On the same day, two of the Unit helicopters operating at Sylhet experienced heavy ground fire but were safely flown back to Kailashahar.

8th Dec – Battle of Sylhet The air superiority gained was used to great advantage. Lt Gen Sagat Singh, Gp Capt Chandan Singh & Sqn Ldr CS Sandhu (CO 110 HU) worked out a contingency plan to operate at night to prevent being spotted by the enemy and reduce the possibility of enemy fire. Handheld battery powered torches were fixed on ground and used as landing aids for helicopters. At midnight, a helicopter airlifted vital equipment and more troops from Kalaura to Sylhet under heavy enemy ground fire, flying three sorties at 0100 hrs. Four more helicopters undertook night SHBO missions and flew 14 sorties airlifting 184 troops and 300 kg of load. Two more helicopters from Aizwal joined the fleet at Kailashahar and undertook SHBO missions to mobilize 95 troops & 9400 kg of load to Sylhet during day time. During these missions all the helicopters came under heavy ground fire and two of them were badly damaged. Though crippled, they managed to fly back to Kailashahar safely.

9th Dec – Sylhet and Raipura This was the last day of Sylhet SHBO. 584 troops and 12,500 kg of load were airlifted to Sylhet. All helicopters moved to Agartala by 1200 hrs. A new task was given to lift troops and load from Brahmanbaria to Raipura. 27 sorties were carried out airlifting 300 troops and 2,200 kg of load. Though ground fire was encountered enroute Raipura, no helicopters were damaged

10th Dec – Raipura The Commander of 4 Corps decided not to stop on the east bank of the Meghna River but to get across if possible to Dacca. The bridge across Meghna was blown up and it was left to the helicopters to heli-bridge a sizeable force across the river. The operation to cross River Meghna began shortly before last light when about 100 troops were ferried to landing pads. By the time the second wave of helicopters arrived it was quite dark but the helicopters landed safely. The movement of troops continued throughout the night and the next day. Though the normal capacity of Mi-4s was to carry 14 troops, the number of troops was increased to 23 men per trip. During the Raipura SHBO, pilots and technicians worked without any relief throughout the day and night and a total of 650 troops and 8200 kg of load were airlifted; while also evacuating 104 casualties in 57 sorties.

11th Dec – Narsinghdi The Vanguards were given the largest SHBO task when eight helicopters of 110 HU were moved to Narsingdi from Brahmanbaria. By 0730 hrs every helicopter had done three sorties each, airlifting 321 troops and 9200 kg of load. After refuelling at Agartala the second wave took off at 0900 Hr and carried out three sorties each airlifting 246 troops and 14600 kg of load. Two more details were flown by the helicopters. In all, a total of 834 troops and 58,600 kg of load were airlifted involving 99 sorties and 65:30 hours of flying. More sorties could have been carried out but the ground element could not cope up with the speed and tempo with which the SHBO was carried out.

12th Dec 35 sorties involving 28:15 hours of flying were carried out at Narsingdi, airlifting 234 troops and 18,520 kg of load. Due to heavy ground fire, there were 30 casualties and one helicopter was sent to Sylhet for their evacuation. One helicopter met with an accident due to fire in air.

13th Dec 282 troops and 14,850 kg of load were airlifted in 30 sorties and 25 hours of flying.

14th Dec 12 helicopters were positioned at Daudkandi at 0730 hrs where they were employed to move a battalion of troops across the River Meghna to Baidya Bazar, barely seven miles from Narayanganj on the outskirts of Dacca. By evening, 810 troops and 22,650 kg of load was airlifted in 79 sorties and 38:20 hours of flying. From the time that the SHBO operations into Narsinghdi began on 10 Dec evening, till the landing of a battalion at Baidya Bazar, pilots had already flown for 36 hours continuously doing a total of 409 sorties. The helilift of troops at these three places– Sylhet, Narsinghdi and Baidya Bazar– was accomplished without a single casualty.

15th Dec SHBO missions continued and 1209 troops and 38,100 kg of load were airlifted to Baidya Bazar involving 121 sorties and 63:20 hours. One aircraft met with an accident but the morale of the Unit continued to be high.

16th Dec Most of the helicopters were taken up for maintenance and inspection. One helicopter which was sent to Sylhet for casualty evacuation airlifted 16 casualties while two other helicopters were utilised for army communication sorties during the day. In the evening five helicopters took part in the Dacca Surrender Ceremony.

MI 4 HELICOPTERS DISGORGING TROOPS IN SYLHET (EAST PAKISTAN) DURING 1971 OPS | 110 HU KEEPING THE JOINTMANSHIP ALIVE DURING THE 1971 OPS EAST OF DHAKA

SHBO and Cas Evac Missions Flown by 110 HU in Dec 1971

Sl No	Mission	Sorties	No of Troops	Total Load (kgs)
1.	SHBO Sylhet	66	1117	22,600
2.	SHBO Raipura	57	650	10,400
3.	SHBO Narsinghdi	164	1350	91,950
4.	SHBO Baidya Bazar	200	2019	60,750
5.	Cas Evac	-	269	-
	Total	**487**	**5136**	**1,85,700** (185.7 T)

As is evident from the accounts listed by various sources, 110 Helicopter Unit was an integral part of the 1971 Operations east of Dacca. Mi-4 helicopters undertook extensive heli-bridging operations in the river dominated terrain of East Pakistan where all the major bridges had been blown up by the enemy. While the initial part of operations commenced by airlift of troops to Sylhet on the 7th, the tempo and risk of operations kept increasing by the day. Mi-4 helicopters were instrumental in the progress of battle from Raipura to Narsingdi, Daudkandi and Baidya Bazar. The enormity of the task undertaken can be gauged from the over 487 sorties that were flown by the Unit in the direct face of enemy fire during these operations.

Snow Tigers: 111 HU 111 HU proved itself during the 1971 war of liberation of East Pakistan. Under the command of Sqn Ldr K C Cariappa, the Unit's helicopters operated from dispersed locations and carried out a variety of tasks, namely rocket attacks, river crossing operations and casevacs. Once again the mighty Mi-4 and the personnel of 111 HU had combined into a hard hitting combat force. With the creation of Bangladesh, the war came to an end leaving behind a host of gallantry awards, stories of courage and memories to evoke pride in generations to come of those glorious days.

105 HU The unit undertook the boldest of tasks, by airlifting 6023 troops and 1,79,160 kg of military hardware around Dacca, during day and night operations under adverse conditions over a territory both unknown and hostile, dogged by deadly, persistent enemy fire. 105 HU also contributed its might in the birth of Bangladesh throughout by its humanitarian endeavour.

Hovering Angels: 115 HU On 15th Sep 1970, 115 HU entered the jet age and was re-equipped with the more sophisticated Allouette III helicopters of French origin. With war clouds hovering over the Eastern front, the unit was tasked with two detachments of one helicopter each at Teliamura in Tripura and Kumbhigram in Southern Assam. It took active part in the Bangladesh operations and acquitted itself commendably. The "Hovering Angles" had a busy time in the operational area, sighting the enemy, searching their hideouts and evacuating casualties. Many recce missions were flown over enemy territory towards the overall operations in then East Pakistan. On 9th Dec 1971, between 1215 hours to 1245 hours, Fg Offr Sidhu as Captain and Fg Offr Sahi as Co-Pilot, with Commander 17 Corps on board, took off from Kumbhigram for an operational recce of an area near Bhairab Bazar in East Pakistan. Near Bhairab Bazaar, the helicopter was fired upon. Pilots took immediate evasive action, but before they could go out of the firing range, the helicopter was hit by bullets. One of the stray bullets pierced through the collective and Sidhu was injured. The aircraft was flown back safely to Agartala by Sahi, the co-Pilot, despite sustaining some damage. The crew displayed exemplary skill and valour and flew the aircraft to safety at Agartala. Fg Offr Sidhu was then evacuated to a hospital. This incident did not deter the Chopper crowd at all and operations continued with the same enthusiasm as before till the detachments returned to base on 18th. During the short span 10 many as 212 sorties were flown covering 153 hours. It should be mentioned here that the unit undertook its first night cross country flight during the war. Night flying in hilly terrain with no navigational aid was an achievement by itself. It also maintained a detachment at Dhaka for six months after cessation of hostilities.

117 HU: Himalayan Dragons The unit was extremely fortunate to be involved in a major operation after its formation in the Indo-Pak war of 1971. During these operations, a detachment of four helicopters carried out 287 sorties in 166:45 hours with 100% serviceability in the eastern theatre. It carried out search and rescue operations, casualty evacuations and communication sorties in East Pakistan. The first IAF aircraft to land in liberated Bangladesh were two helicopters of this unit, which landed at Dacca on 16th Dec and brought Lt Gen Niazi, Commander Pak Army in

East Pakistan and Maj Gen Rao Farman Ali, PAF Commander in EAC as POWs from Dacca to the Eastern Command HQ at Red Road on 20th Dec. It also provided airlift to the President, Prime Minister, Defence Minister and the services chiefs on various occasions during the operations.

The Knights: 109 HU The inevitable happened in the first week of Dec 1971. One Zero Nine not only rose to the occasion but surpassed all expectations. It was the first to land at Jaurian and lift out the first batch of casualties. The unit flew a total of 315 hours during the operations and lifted out a total of 468 casualties directly from battle zones under the enemy fire besides doing hundreds of sorties for air logistics support and recce. The mementos presented to this unit stand silent yet eloquent testimonials of the gratitude and confidence reposed by the Indian Army.

104 HU: The 1971 war was a glorious chapter in the history of the unit. Rumblings of war with Pakistan were felt as early as Feb 71 when an Indian Airlines aircraft was hijacked to Lahore. For 104 HU, the war virtually started in August when it was called upon to maintain detachments at Jullundur, Amritsar, Halwara, Pathankot and Sirsa. Thus, most of the aircrew though unarmed, took active participation in the operation. The primary task assigned was to undertake aerial reconnaissance, communication and casevac missions in support of the Indian Army. True to the tradition of IAF, the pilots vied with one another to undertake missions. The zeal and enthusiasm shown by all the pilots was to be seen to be believed. Some of our pilots had a very narrow escape when they were spotted by enemy fighters but managed to remain evasive. During these two weeks, a total of 268 hours in 433 sorties were flown, and Sqn Ldr SM Hundiwala, the then Commanding Officer, was awarded Bar to Vayu Sena Medal for distinguished service of exceptional order.

114 HU: Siachen Pioneers During the operations, the unit was allotted the task of casualty evacuation, communication and reconnaissance of entire Jammu and Kashmir sector. Many reconnaissance sorties were flown to detect infiltrators during the actual operations. One helicopter while carrying out casualty evacuation in Kargil sector met with a fatal accident while taking evasive action from a Pakistani fighter aircraft. A number of sorties were carried out for evacuation of Indian Army personnel from the battle front. These casevacs were carried out on top priority from the battle zone.

CITATIONS: 1971

Maha Vir Chakra

Group Captain Chandan Singh, AVSM, VrC (3460), Flying Branch (Pilot).

Group Captain Chandan Singh was the Officer Commanding of an Air Force Station in Assam. During the recent conflict with Pakistan, he was in the forefront of the air operations conducted for the liberation of Bangladesh. He was responsible for the planning and execution of the special helicopter operations to airlift two companies of troops to the Sylhet area. When it became necessary to overcome the obstacles in the advance of the Army towards Dacca, he planned and executed the move of nearly 3000 troops and 40 tons of equipment and heavy guns with the extremely limited helicopter force at his disposal. This operation entailed landing the troops and equipment very nearly heavily defended areas by night. Prior to each mission, he personally carried out reconnaissance in the face of heavy enemy fire. On the night of 7th/8th December he flew eight missions, deep into enemy territory, to supervise the progress of the helicopter airlift and to guide his pilots who were facing heavy opposition from ground fire. Later he undertook a further 18 missions in the same operation, always leading the landings at new places. On many occasions, the helicopter was hit by ground fire, but this did not deter him from further missions. The success of this major airborne operation contributed significantly to the fall of Dacca and the capitulation of the Pakistan armed forces in East Pakistan.

Vir Chakra

Flt Lt Chander Mohan Singla

During the Indo-Pak war in 1971, Flt Lt Singla served with a helicopter unit in the Eastern Sector. One of his vital tasks was to convert some foreign pilots (with transport, bomber and fighter backgrounds) to Chetak helicopter by day and night. He completed the assigned task by the deadline under arduous conditions. Subsequently, his operational task included the exploitation of armed Chetak helicopter by day & night. The operations matured into the formation & operation of helicopters as a self contained & mobile hitting outfit, supporting our advancing ground troops. A remarkable incident took place on the night of 7th/8th December 1971. A special heliborne operation was under way to Sylhet using Mi-8 helicopters. Singla was given the task of providing armed cover for these operations. During this mission, the enemy directed intense ground fire at his Chetak helicopter. It was hit at eight places. he pressed on regardless, and gave effective air cover for 4 ½ hours, during which 35 rockets were fired at ground targets with exceptional accuracy. Thereafter, he continued to operate on equally hazardous missions every night from 7 Dec to 12 Dec at various places in enemy territory & deep behind enemy lines. Flt Lt Singla was awarded the Vir Chakra for singular display of courage, gallantry, professional skill and devotion to duty.

Flight Lieutenant Pushp Kumar Vaid (6892) Flying (Pilot)

During the operations against Pakistan in December 1971, Flight Lieutenant Pushap Kumar Valid was serving with a helicopter unit deployed in the Eastern sector. On one occasion it was decided to transport an infantry element to the sector by air. Flt Lt Pushap Kumar Vaid, knowing full well that the helicopter landing area would not be adequately lit and that he would come under heavy enemy fire on landing, volunteered to undertake this mission, and successfully completed the task. Subsequently, he flew 34 more hazardous missions deep behind the enemy lines.

Flight Lieutenant Mahabir Prasad Premi (8378), Flying (Pilot)

During the operations against Pakistan in December 1971, Flight Lieutenant Mahabir Prasad Premi was commanding a helicopter detachment in the Rajasthan Sector. On 6th December, while engaged in casualty evacuation he discovered an unexploded bomb lying near his helicopter. Realising that if the bomb exploded, the helicopter would be damaged; he decided to take off. After he had started up, an air raid alert was given over the airfield, but in spite of it he took off and landed in a safer area. Immediately after this, the bomb exploded but by his timely action the helicopter was saved. On 11th December 1971, he evacuated two casualties from an area which was under enemy air attack.

Squadron Leader Charanjit Singh Sandhu (5591), Flying (Pilot)

During the operations against Pakistan in December 1971, Squadron Leader Charanjit Singh Sandhu was in command of a helicopter unit. On the night of the 7th/8th December 1971, he personally led a special heliborne operation deep behind the enemy positions in the Eastern Sector. During the first mission, his force came under enemy fire, but he successfully carried out the landings of his force. The same night, he flew an additional five similar missions. During the operations, he led a total of 34 such hazardous missions.

Flying Officer Bartan Ramesh (10948), Flying (Pilot)

During the operations against Pakistan in December 1971, Flying Officer Bartan Ramesh was serving with a helicopter unit. He carried out forty-five successful operational missions, some of which were in the heart of the battle area. During two of these missions he, with complete disregard to his personal safety, evacuated Army battle casualties from forward helipads in the Uri Sector in the face of constant shelling by the enemy.

Flying Officer Sukhdev Singh Dhillon (11378), Flying (Pilot)

During the operations against Pakistan in December 1971, Flying Officer Sukhdev Singh, Dhillon flew as a pilot in a helicopter unit and single-handedly evacuated 87 battle casualties from the most difficult and hazardous terrain of the Kargil Sector in complete disregard to his personal safety. He carried out these evacuation missions in the face of heavy ground and air opposition.

Flight Lieutenant Parvez Rustam Jamasji (9834), Flying (Pilot)

During the operations against Pakistan in December 1971, Flight Lieutenant Parvez Rustom Jamasji was serving with a helicopter squadron. The helicopter flown by him by him was attacked twice by machine guns and mortars, but he fulfilled the mission assigned to him and brought back his aircraft to the base. On one occasion, his helicopter had engine failure over enemy position, but he brought it safely to a post within our territory.

Squadron Leader Sanjay Kumar Choudhary (5863), Flying (Pilot)

During the operations against Pakistan in December 1971, he served in the Eastern sector. On the night of the 7th/8th Dec, a special heliborne operation was undertaken for landing a battalion at Sylhet which was at that time under enemy occupation. Squadron Leader Choudhary volunteered and landed with the first wave of troops to personally hold up and supervise the lighting of flares. During the process, the helicopter landing area came under heavy enemy gun fire. Regardless of own safety, with the help of one jawan, he made efforts to light up to landing zone. Shortly thereafter the jawan was killed by enemy fire. Undeterred, he single-handedly lit the flare and kept it lit the whole night. This enabled the heliborne operation to continue the whole night and its successful completion.

Flight Lieutenant Harjit Singh Bedi (7008) Flying Pilot

During the December 1971 operations, he flew twenty six operational missions in the form of reconnaissance emergency supply and casualty evacuation from one of forward airfields in Rajasthan Sector. He carried out three missions despite bombing and strafing of the areas by enemy aircraft. On 14th December 1971, while he was going to land at a helipad in the forward area for evacuation of casualties, four enemy aircrafts attacked in the vicinity. Despite these attacks he landed his helicopter in such a manner that the enemy aircraft could not spot it. He thus not only saved the helicopter, but also remained with it and flew it back to base.

Flight Lieutenant Prakash Nath Sharma (9412) Flying (Pilot)

During the Indo-Pak operations in December 1971, he as pilot of a Helicopter Unit operated from forward helipad in support of our ground forces. He flew fifty missions deep inside enemy territory carrying food, water and ammunition for our ground forces in Nayachor, Chhachor and Khinsar areas and bringing back battle casualties. Some time he had to land at completely unprepared desert surface, well within the range of enemy fire. On 11th December 1971, while returning from a mission, he was attacked by four enemy aircraft about six miles from Nayachor. Realising the futility of escape, he immediately landed his helicopter between two sand-dunes and ordered his crew to take cover under the thick dust kicked up by the rotor. Each enemy aircraft strafed the helicopter and the crew by turn and one of them dropped a 600 lbs bomb also, but their attacks failed to find the mark. The only damage caused was that the perspex of the helicopter get shattered and the reductor bay cowling was blown apart. Undeterred, he took off again and landed at a forward helipad and thus not only saved the helicopter but also the lives of the crew members.

Flight Lieutenant Ravindra Vikram Singh (9524) Flying (Pilot)

During the Indo-Pak operations, he led a detachment of helicopters in the Eastern Sector. He displayed great devotion to duty and undertook hazardous tasks over hostile and difficult terrain. By his inspired leadership he moulded his team into a highly organized and effective force, capable of giving their best at all times. On 16th December 1971, Flight Lieutenant Singh as captain of a helicopter was conveying the C-in-C and Chief of Staff of Bangladesh with other passengers to Sylhet. While flying to Sylhet at a height of one km, his aircraft was hit by enemy ground fire at 1720 hrs, in which one bullet, after piercing the cabin and injuring the Chief of Staff of Bangladesh, hit an oil pipe, thus resulting in leak of hot oil into the cabin. Flight Lieutenant Ravindra Vikram Singh, realizing that his aircraft was hit, pulled up immediately to height of 1.5 km, and by this time the reductor oil pressure had started dropping rapidly. It was by then dark and returning back to base or to any other airfield was impossible. At this stage he displayed great professional skill, brought the aircraft out of enemy area, selected a suitable field at Fenchuganj after approximately 15 minutes flying under trying circumstances in complete darkness and made a successful landing saving the aircraft and lives of C-in-C and chief of Staff of

Bangladesh along with other passengers and crew. Earlier on 8th December 1971, Flight Lieutenant Singh displayed exceptional qualities of airmanship and brought aircraft back safely when it was hit just six inches behind the co-pilot seat, twice on main rotor blades and once on main gear box cowling by enemy ground fire.

Flight Lieutenant Ramesh Thakurdas Chandani (9797) Flying (Pilot)

During the December 1971 operations against Pakistan, he carried out twenty six sorties in Jammu and Kashmir sector. On 14th December 1971, at about 1600 hours, he was flying back from Nainakot which was an improvised helipad in captured enemy territory and which was under constant threat of enemy air and ground attack. On the way, his co-pilot drew his attention to one of our aircraft which had been shot down and was on fire. He immediately landed near the wreckage and rescued an Army Officer who was the only living occupant of the aircraft and thus saved his life.

202634 Warrant Officer Koshy Thomas VM Flight Engineer

He was serving with a helicopter unit in the Eastern Sector during the Indo-Pak conflict in December 1971. His unit was engaged in airlifting troops/armament and casualties. He always volunteered to fly in all types of sorties, both by day and at night, without any regard to this personal comfort. Throughout the special heliborne operations at Sylhet, his helicopter was under heavy enemy ground fire. With complete disregard to his personal safety, Warrant Officer Koshy Thomas courageously got out of his helicopter in enemy territory and supervised the unloading, thus enabling quick and successful completion of the task. He carried out 124 sorties which included 26 sorties by night.

The Birth of Bangladesh

CHAPTER VI

Operations in Sri Lanka

Introduction In the mid-Eighties, new acquisitions of Mi-25 attack helicopters and the powerful Mi-17s had considerably boosted helicopter capabilities of the IAF. However, the proof of the pudding lies in the eating – and this opportunity dawned with the Indian Peace Keeping Force (IPKF) in Sri Lanka in the late-Eighties. Offensive and innovative roles drawn on the drawing boards saw fruition from 1987. It was an entirely new realm of joint-fighting models that emerged in Op-Pawan. The joint-structures with air force elements and commanders interspersed with the army formation brought about dramatic results. The sheer volume of effort of the helicopter and transport aircraft fleet towards the support of the Indian Army is humungous – it has not been matched by any country except for the US in Iraq and Afghanistan.

The concept of airpower being handled by the air force experts yet integrating fully in the overall campaign was a redemption of all that is taught and ingrained in various war colleges and joint institutions. While the politics in the Sri Lanka campaign takes centre-stage in all analysis, the role of the military was quite commendable considering the numerous flip-flops and see-saw of international politics. By the time the IPKF pulled out, the LTTE was on its knee and it was just a matter of time before end-objectives could have been met. One of the highlights of IAF support was effective changes in tactics and strategy to adapt to all types of Indian Army offensives – both large and small scale operations. Night operations of every kind without night vision goggles were another feather in the cap. The living conditions and environment on ground and in the air were highly stressful. Movies made on Vietnam by Americans suddenly started making sense! A great achievement of the helicopter fleet was the little or no collateral damage in the entire campaign – reference all campaigns of the US/NATO or even by the Sri Lankan armed forces since then. This showed great maturity and understanding of root issues in such conflicts.

This chapter details the herculean effort that the IAF helicopters put in. A similar work was done by the transport fleet. While sketchy mentions are made in many chronicles by the Indian Army, it is a moot point that Op-Pawan may not have moved anywhere without the unstinted support of the IAF. Personal accounts and citations at the end clearly bring out the dedication of helicopter aircrew towards the task of supporting IPKF. Pilots repeatedly went into combat zones to ensure that the support given was empanelled in the annals of military helicopter legacies. The Jaffna University SHBO is just one such example. Initially blamed for the fiasco, the gallant aircrew were feted and rewarded with gallantry awards once it was realised that the fault in planning and assessment lay with the Indian Army. The aircrew just did what they were trained for– give their heart and soul to the job in hand.

1971: First Forays into Ceylon India's involvement in Sri Lanka in times of peace and war has been regular since the countries gained independence from British rule. The Indian Peace Keeping Force (IPKF) deployment in late 1980s is an important chapter in India's military history. But few recall India's aid to Ceylon (as Sri Lanka was then known) to cope with the violent insurgency launched by JVP in 1971.

The Janatha Vimukthi Peramuna, an extremist-leftist Sinhalese organisation attempted to seize power through force in April 1971. Although not a trained military force, it's cadres seized and captured large areas in southern and central provinces. The group had been active for some years, but due to growing action by the police forces it decided to launch the full blown insurgency on 5th April. Rebel groups armed with shotguns, bombs, and Molotov cocktails launched simultaneous attacks against 74 police stations around the island and cut power to major urban areas. Royal Ceylon Air Force base at Ekala was also attacked and other attacks were most successful in the south. By 10th April, rebels had taken control of Matara District and the city of Ambalangoda in Galle, and came close to capturing the remaining areas of Southern Province.

The government under Prime Minister Srimavo Bandaranaike was not prepared for the crisis that confronted it. Sri Lanka's all-volunteer army had no combat experience since World War II and no training in counter-insurgency. Although the police were able to defend some areas unassisted, in many places the government deployed personnel from all three services in a ground force capacity. RCAF's No. 4 Squadron – equipped with just three Bell 206A Jet Ranger helicopters at Colombo – delivered relief supplies to besieged police stations and military outposts. The squadron could muster

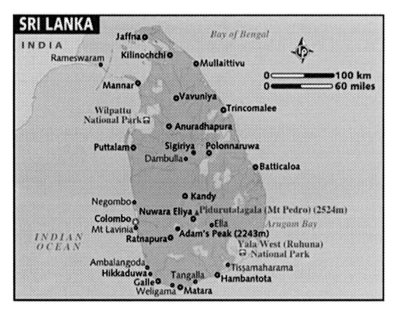

only 22 hours of flying in the first five days. Joint service patrols drove the insurgents out of urban areas and into the countryside.

The authorities were caught off-guard by the scale of the uprising and were forced to call on India for armed help. Srimavo Bandaranaike sent a distress signal to Indira Gandhi, the prime minister of India. But as the telecommunication system in Ceylon by that time had been damaged, New Delhi managed only to receive a garbled cable from the Ceylon prime minister. As assistance was not forthcoming from India, KPS Menon the Indian High Commissioner, was sent to New Delhi to personally convey the distress call. Indira Gandhi hurriedly summoned her cabinet to discuss the distress call. The government answered the call for help with the deployment of all three Services in a joint-operation.

Indian Deployment Lieutenant-General GG Bevoor commanding the Indian Army's Southern Command was alerted. Troops were urgently airlifted from Bangalore and Madras (Chennai), to the Ceylonese air force base in Katunayake. Five Allouette III/Chetak helicopters from 104 Helicopter Unit were deployed to Ceylon. The unit had operated in Ceylon previously; during 1957-58 floods in Ceylon, two Bell Helicopters from 104 Helicopter Flight (now 104 (H) Sqn) were airlifted to carry out relief operations from 28 Dec 1957 to 10 Jan 1958. It was based at Hindan Air Force Station at that time and commanded by Sqn Ldr SM Hundiwala, VM & Bar. 104 HU chalked up 573 flying hours in 1122 sorties in these operations in 1971. Ultimately, it took two weeks of fighting for the government to suppress the JVP rebellion with considerable efforts from the Indian armed forces. The Ceylon peacekeeping operation was independent India's first overseas operation under own command (other foreign deployments were under UN). Little details are available since the event was eclipsed by the much larger scale Bangladesh Liberation War of 1971.

OPERATION PAWAN: IPKF SRI LANKA

An Overview

In 1986, the LTTE, just like an organized army, was a five command organization – one each for Jaffna, Vavuniya, Mannar, Trincomalee and Batticaloa. Each of these commands was divided into a 'political' wing and a 'military' wing. The political wing administered LTTE controlled area while the military wing was responsible for offensive operations against the government and other rival groups. Jaffna remained the strong hold; but LTTE displayed great flexibility in deploying troops wherever the need arose.

As part of operations Poo-Malai, which means 'a garland of flowers' in Tamil, a force package of five An-32s escorted by four Mirage-2000s, dropped over 22 tons of supplies in Jaffna. The fighter escorts were there for the obvious reason – to ward off any attempt by Sri Lanka to intercept the An-32s. Op Poo-Malai demonstrated India's will and ability to intervene and was enough to shake the confidence of the Sri Lankan President Jayawardene. This action left them with no choice but to accept the Indian Government's diplomatic advances and open up formal as well as informal channels of negotiations.

IPKF Preparations The GOC-in-C Southern Command was made the overall force commander with the following main forces under him:

- 4 Inf Div (2 Bdes only)
- 18 (Independent) Inf Bde
- 36 Inf Div
- 54 Inf div
- 57 Inf Div
- 340 (Independent) Inf Bde
- 664 AOP Sqn (Chetaks & Cheetahs)
- 19 Sqn AF   – An-32s
- 109 HU AF   – Mi-8s
- 119 HU AF   – Mi-8s
- 125 (H) Sqn   – Mi-25s

Apart from the Army and Air Force components as above, the Indian Navy contributed with two naval helicopter sqns -310 Sqn (Breguet Alizes) and 321 Sqn (Chetaks) – MARCOs and small patrol boats monitoring Palk Strait and Gulf of Mannar. CRPF and Indian Coast Guard were the two paramilitary forces that lent a helping hand. The aim of the Indian Peace Keeping Force (IPKF) was to defeat the LTTE completely, or, at least compel it to fully accept the Indo-Sri Lanka Accord (ISLA). In spite of the fall of Jaffna, this did not come about as the LTTE reverted to its best fighting technique – the guerrilla war. For two years, the IPKF fought the war with LTTE but had to withdraw in 1990 without achieving the main political aim.

After the initial hectic but peaceful deployment, eventually the IPKF had to be involved in direct fighting with LTTE. On 3rd Oct 1987, the Sri Lankan Navy captured two LTTE commanders (Kumappa of Jaffna and Pulendran of Trincomalee) and 15 others; who were smuggling arms into Sri Lanka. Transferring them to Colombo was the new point of conflict. The IPKF and the Indian High Commission advised against the shifting and the captured men threatened to commit suicide if moved to Colombo. All negotiations failed and the captured cadre did commit suicide (only two could be saved). The LTTE retaliated by killing around 200 civilians and eight soldiers. This episode is what can be called the final nail in the coffin of the ISLA (Indo-Sri Lanka Accord) and the decision for eventual action against the LTTE starting 7th Oct. Almost simultaneously, the LTTE carried out its first direct attack on IPKF; which (naturally) retaliated. The war had begun.

The assault by IPKF commenced on 9th Oct, with probing raids on camps at Jaffna to test the fighting capabilities of LTTE and to demonstrate that IPKF was serious about its job. 1 Maratha LI attacked radio and TV broadcasting centres and printing press of LTTE with an aim to scuttle their communications. After the 'softening up', the main assault of Op Pawan began on 11th Oct when 4 Mahar and 12 Grenadiers attacked Jaffna from the elephant pass. Simultaneously, 5 Mahar and 8 Mahar attacked from the North. The IPKF also attempted to get to the LTTE high command with the helicopter raid on Jaffna University. Had it been executed with the adequate resources, this raid might have been successful too. Leaked information, however landed them straight into the enemy's jaws and as a result only one jawan survived the battle

Guerrilla Warfare Many LTTE fighters escaped Jaffna and avoided capture but the fall of Jaffna severely crippled their operations. LTTE's communication, which was radio based, was effectively targeted by IPKF after the capture of Jaffna. The logistics of LTTE, ammunition to be more specific, suffered a huge set back. During this time, IPKF operations were sub divided into four sectors namely, north or Jaffna sector, east or

Trincomalee sector, south east or Batticaloa sector and central or Vavuniya sector.

One of the fiercest battle occurred in the beginning of March 1989. After the provincial council elections, the IPKF began operation Baaz aimed at bringing the LTTE to the talking table. The freshly inducted 6/8 GR was on a two day patrol that eventually transformed into a two week battle. With limited experience of search at planned road blocks only, 6/8 GR found hope of some action in this patrol. Wading through thick jungles close to Nayaru Lagoon, the lead platoon ran into an ambush losing five jawans in the opening burst of fire. The Gorkhas fought back but there was no way of gauging the number of LTTE casualties. The experience of previous encounters had shown that such major strong holds of LTTE were often guarded in two tiers over 10 km and Col Bakshi, the commander of the 6/8 GR opined that this resistance was somewhat akin to the outer ring. This belief resulted in a vigorous assault from the Gorkhas. Col Bakshi also asked for reinforcements. By the nightfall, the Gorkhas ran out of ammunition and drew out their trusted Khukris. Col Bakshi was fatally wounded. The reinforcements, an infantry company, dropped by IAF helicopters, found themselves on a heavily booby-trapped trail. Soon, five battalions of IPKF marched to the lagoon. Over five days, about 70 LTTE personnel were killed. But a majority of them still managed to escape and almost all of the commanders of LTTE managed to give the IPKF a slip. By the end of the two weeks, an entire division strength of IPKF troops was at this battle field. IAF Chetaks guiding the Mi-25s and the Mi-8s dropping troops, ammunition, equipment and food supplies, were largely responsible for the large scale mobilization. Mi-25s carried out a large number of attack sorties while the Chetaks and Mi-8s looked after the casevac.

Many such operations caused the LTTE to lose ground and run to hide deeper in the jungles. Operations like Steel Gray, Lilac, Swordfish, Rolling

Trumpets and Kiranthi were all conducted with the same aim. Having cleared the Batticaloa region, search and comb operations like Op Blooming Tulip were conducted. Operation Blooming Tulip forced the LTTE out of their 10-village base called Beirut.

Reasons for the Debacle A number of analysts have brought out that the IPKF followed the wrong doctrine in this war with the Tigers. What was followed was based on operations of a large unit (usually division strength) involving close supervision and direction from higher command. The challenges of unconventional tactics were not fully understood and reliance on conventional ones would eventually cost us dear. Apart from command & control (due to ad-hoc attachment of units to IPKF) and poor intelligence, one major problem faced by the IPKF was its strategy to stick to major roads for movement. It involved moving along well established roads till contact with LTTE forces was made and then fighting them. For an enemy as elusive as the Tigers, this proved highly ineffective. To add to the woes of IPKF, this allowed the LTTE to easily lay IEDs and ambushes and delay the advance. The IPKF preference for large scale operations like in the conventional warfare was evident in the combing operations.

Many reports were received stating that Prabhakaran was sighted in a particular area. The resulting search operations were as large as two Bdes, a mechanised infantry company, a tank squadron and helicopter gunships to search 150 square kilometres. Somewhere in Feb 1988, an even larger force was used to search for him south of Jaffna. These methods obviously failed as he was highly proficient in eluding capture. Operation Viraat was launched in May 1988. It was a massive force of 15000 infantry troops, armoured and Para forces. While carrying out a combing search, the IPKF came across fortified LTTE defences including some concrete bunkers. The search operations were very often ineffective and the LTTE cadre would employ hit and run tactics. The only IPKF forces which managed some success were 9 Para who employed tactics similar to those of the LTTE. The success of Op Merry Lane proved that such tactics were better. The SF realized the importance of intelligence, especially regarding jungle routes. Just like the political and military wings of LTTE, the SF operated two elements for intelligence gathering and strike arm. While difficult and time consuming, intelligence gathering was essential to the success of all IPKF operations – and the SF commander knew it well.

The relationship of the Indian diplomats in the Indian High Commission in Sri Lanka and the fighting forces of IPKF were not helping the situation. The officers of IPKF complained of interference on part of the High Commission while the diplomats accused the IPKF of not fully understanding the mission requirement and non-cooperation. The forces were pretty vocal about the perceived intervention and the diplomats were equally open in letting know the IPKF that the conflict was as much political as military. Indian government's policies towards LTTE made IPKF officers think that war with LTTE was not imminent. Added to that, the location of

IPKF HQ in the LTTE dominated Jaffna made it mandatory for IPKF to accept LTTE demands. Open admiration of LTTE's discipline and devotion in the views of the IPKF cadre was a mere catalyst. All the above combined and the result was that many senior IPKF officers seemed willing to meet the increasing demands of LTTE rather than prepare for a war with them.

Integrated Joint Operations

On 23rd Oct 1987, a detachment of Mi-25 gunships from 125 HU from Pathankot was inducted into Jaffna. This detachment of three aircraft led by Wg Cdr SC Malhan had been standing-by in Sulur a week earlier, for possible induction into the island. On 25th, a Mi-25 carried out its first live mission, led by S Hooda. The target was Jaffna jetty. The armament expended was 128 × 57 mm rockets and 328 × 12.7 mm front-gun rounds. The results were almost immediately apparent. The LTTE was forced to stay away from roads and waterways and restricted their movements to night only. Following this, a number of search and strike as well as pre-planned missions were mounted by the Mi-25s against the LTTE. The varieties of targets engaged by Mi-25s were: small bands of militants; moving vans/tractors and motor cycles; speed boats in lagoons/coastal waters; pre planned missions against huts/bunkers/jetties/jungle hideouts. Most of the attacks carried out by the Mi-25s proved effective and accurate. Some of the important engagements during this phase are given below.

Date	Armament	Target	Result
25 Oct 1987	128×57 mm RP	• Ammo dump near jetty	• Ammo dump near jetty seen exploding
	328×12.7 mm FG	• 3 Vans on jetty	• Van destroyed
			• Estimated 5 men killed and some seen jumping into the water
25th Oct	64×57 mm RP 12.7 mm FG	• Warehouse in Jaffna town near jetty	• On being hit the warehouse exploded – suspected fuel dump. • 2-3 boats tied alongside warehouse were also damaged
26th Oct	Unknown	• LTTE HQ in Jaffna town	• HQ building destroyed. Identified and confirmed by own troops
27th Oct	16×RPs 400×FG	• LTTE bus near Pt Pedro	• Confirmed destroyed
28th Oct	64 RPs/ 260 rounds	• 30 Barracks at Chavaka-cheri stronghold	• Target destroyed • One truck parked outside destroyed

On 31st, Jaffna town was captured by the IPKF. In addition to the operational commitments, the medium of air was very extensively utilised during November and December for additional induction, air maintenance, airlift of specialist equipment/personnel for reconstruction and for large scale regrouping/redeployment of various units/formations. The potent combination of IAF gunships and tactical helicopters (Mi-25/Mi-8) contributed to a large extent in forcing the LTTE to abandon direct confrontation and switch over to guerrilla tactics.

Employment of Mi-25: Phase–II Employment of Mi-25s changed significantly during this phase. The classical Search and Strike missions of Phase I could not be executed because of serious identification problems. A total of three Mi-25s of 125 HU were available out of which any two were serviceable at a given time. The Mi-25s were employed in the following roles:

- Suppressive fire in general area of the intended landing zones of own tactical helicopters.
- Area bombing/strafing of known jungle hideouts. (This naturally had to be restricted to areas where there were no civilian habitats.)
- Armed escorts/CAPS to tactical helicopters engaged in SHBOs.
- CAS of own troops from 'Op Call' position (on ground at forward helipads).

Op Checkmate This was by far the largest combined operations mounted by IPKF during Phase II of its operations in Sri Lanka. Op Checkmate commenced on 17th Jun 88 in the area of responsibility of 4 Div (Vavuniya sector). Due to the distinct changes in the pattern of Ops, the whole operation was divided into three phases. Thus the period 17th Jun to 5th Jul 1988 was called Phase–I (Checkmate-I), 5th to 17th Jul was called Phase–II and 18th Jul to 15 sep 1988 was Phase–III (Checkmate–III). The hostile terrain was made worse due to non-availability of accurate maps for the ground forces as well as the aircrew. This posed serious problems of navigation and target acquisition for helicopter crew. The only feature available in this vast continuous and featureless canopy of jungle tops was a

single north-south railway line running between Vavuniya and Killinochi. At the end of May/Jun 1988, the modus operandi of the militant had settled into a set pattern of:

- ambushes close to Indian Army posts as troops would tend to be off-guard.
- avoiding villages. They would visit villages for provisions/collection of taxes and would return to their jungle hideouts.
- not staying at one hideout for too long. They continuously changed their locations, which meant a large number of camps spread over vast areas of the jungles.

Our own concept of operations was to: continue to maintain pressure; dominate areas known to be frequented by militants; lay ambushes; prevent coastal hopping; increase psychological operations. During Phase II & III, planned combined operations were mounted in Alampil and Nithikaikulum to ferry out the militants. The tactics employed were standard anti-guerrilla tactics. The time frame for their execution was of vital importance wherein lay the value of the air operations. The Mi-8 was employed in: induction/de-induction of troops; redeployment of troops; resupply of arms/ammunition, including artillery ammunition; resupply of food and water in some cases; casevac and medevac. The total air effort by Mi-8s during Op Checkmate was: 1514 sorties /408 hours of operational task; 460 tons load lifted; and, 9434 troops moved. The period of ninety one days, while 'Op checkmate' was in progress was the heaviest in terms of air support requirements. Of the total air effort of 1514 sorties including positioning/return sorties after induction of troops, the actual load carrying sorties were 810 and the remaining 704 were empty or near empty; though all efforts were made to combine such sorties with air maintenance/re-supply sorties. As the total Mi-8 effort was very large the utilisation of this fleet for the entire Phase II is analysed as follows:

- Load carried against total sorties (1514) – 0.43 tons/sortie
- Load carried against total hours flown (408:00) – 3.44 tons/hr
- Load carried in load carrying sorties only (810) – 1.73 tons/sortie
- Total load carried against hours flown above – 6.43 tons/hr
- Average effort/day – 16.64 sorties/day (4:30 hrs/day)
- Average load lifted/day – 15.43 tonnes/day)

Checkmate I & II The Phase I of Op checkmate was actually a series of SHBOs which were related to immediate reaction to information received from various sources. By the third week of June, it was certain that a large camp/group of camps of the LTTE existed in the jungles east of

Nithikaikulum, covering an area between Nayaryu Lagoon in the north and Kokolai lagoon in the south. In order to confirm the presence of such camps, a special mission was launched from Vavuniya on 23rd Jun 1988. A single Mi-8 escorted by a Mi-25 was briefed to winch down commandos into a jungle clearing near the ruins of an old Portuguese fort. After carrying out a dummy winching operation and while actually reeling down a combatant, militants suddenly emerged into the clearing and opened up with automatic fire. The escorting Mi-25 brought immediate fire to bear on the spot within 30 seconds and suppressed further reaction from the militants killing some ten to twelve of them. The leader of the commando mission, Capt Harpal Singh was killed by a burst which pierced the Mi-8 floor below him. The co-pilot, the flight gunner and a soldier sustained minor injuries. The helicopter received about 14 hits, causing damage to its structure, fire detection/extinguishing system, control rod and the left engine. The captain and his crew handled the situation admirably and recovered their crippled machine safely at Mullaitivu, about five minutes flying time away.

The incident at 0830 hrs confirmed the presence of militants in the area. By 1400 hrs, three companies had been air inducted into Nithikaikulum and by 1730 hrs, TAC HQ of 7 Bde was helilifted from Mullaitivu and Kokkuloduvar to counter the presence of the militants. The next day, additional troops were also helilifted between the afternoons of 23rd and 24th June. At dusk on 23rd, two Mi-25 missions were launched for bombing and rocket attacks. On 24th, two more Mi-25 missions were launched before our own troops started moving in. As part of Op Checkmate a special operation – Op Nesco was launched by the Para Commandos which pushed the militants further south-east into the Nithikaikulum jungle. During the week of 25th Jun to 1st July, a total of 543 Para commandos (approx one combat battalion) were moved by air into the jungles between Nedukeni and Tanniyuthu area in support of Op Nesco. 33 tons of ration/cooked food and artillery ammunition was delivered to the commandos. The total effort was 108 sorties in 34 hours. Alongside the operation, heavy artillery strikes were brought to bear on the likely locations of the militant camps around Nithikaikulum. 500 × 57 mm rockets and 4×250 Kg bombs were delivered by Mi-25s while 4×AS-II missiles were fired by an ATGM Chetak.

The next fortnight (2nd to 15th Jul) saw an intensification of our own operations in Nithikaikulum area. Once again in order not to lose any time, the entire redeployment of troops and their resupply was conducted by air. During these 14 days, 265 sorties were flown, lifting a record 1979 troops and nearly 100 tons of load. Mi-25s delivered 811 × 12.7 mm rounds, 506 × 57 mm rockets and 8 × 250 Kg bombs as suppressive fire.

Checkmate III The week of 22nd to 29th July saw severe clashes as our own troops started to tighten their cordon. To prevent escape and maintain maximum pressure, 54 Div was ordered to mount a combined operation in Vishwamadu jungles east of Kilinochi. Op-Cockerel was launched on 28th Jul 1988. For the first time Mi-8s were used in an armed role. It was also the first time that the air to ground strikes by Mi-25s and Mi-8s were closely coordinated with arty and mortar fire. Detailed and exhaustive briefings were held at HQ 54 Div to ensure optimum utilisation. An airborne fire control and observation post was established in AOP Cheetah to control and direct fire. The effort for Op Cockerel was 163 sorties in airlifting 670 troops and 61 tons of load. 20 casualties were evacuated during this week. As part of Op Checkmate-III, one of the teams of 9 Para Commando had moved to a position which was totally hidden by a thick canopy of tall trees except for a small opening. There was no way this team could be given supplies except by Chetak which had obvious limitations of load carrying capacity. A Chetak deployed team supervised the increasing of a small clearing to permit a Mi-8 landing. One hour later, Flt Lt Gill took off from Nithikaikulum with one ton supplies and successfully executed a landing in the clearing. A finer example of Army/Air Force cooperation would be difficult to find.

SAM-7 Discovery On 1st Aug 1988, one of the captured LTTE soldiers during his interrogation confirmed that he knew the location of one of the two SAM-7 missiles acquired by LTTE. A Mi-8 was immediately launched carrying a troop of 9 Para along with the LTTE prisoner. It landed near a house indicated by the prisoner and a locally fabricated wooden box containing one non-operational SAM-7 was dug out from the back. The missile was probably supplied by Libya.

A Daring Cas Evac An urgent demand was received in the afternoon of 18th Aug 1988 for evacuation of four very serious casualties of 1 Para deployed in Nithikaikulum jungles. Since there was no clearing fit enough for a Mi-8 to land in, it was decided that the casualties had to be winched up. The visibility, light conditions, factors of terrain and possible presence of militants made the decision difficult. The helicopter reached the site almost at dusk after picking up an ALO and two armed commandos as escorts cum lookouts. Three of the casualties were winched up within twenty minutes. After the third winching operation, the captain of aircraft decided to climb up and cool the engine as its temperature had risen excessively because of the continuous hover for twenty five minutes. The fourth casualty, who had serious shoulder and chest injuries was also winched up eventually and all four flown to Trincomalee.

The Nithikaikulum Ops (12-21 Aug 1988) With confirmation of the presence of militants (estimated 100-150) in the Nithikaikulum area, every effort had to be made to encircle them. The only way was to increase the troop strength so that an effective cordon could be established. With this in view, 10 Para and 1/1 GR were moved by air to Nithikaikulum. The air effort involved for this task was 88 sorties, lifting 668 troops and 12 tons of stores in a 24 hour period. On 13th Aug alone, the total air effort was 73 sorties/28:25 hrs. The armed tasks executed in this period were:

- Mi-25: 33 sorties for armed strikes/escort missions, delivering 571 rockets and 36 × 250 Kg bombs.
- Armed Mi-8: 27 sorties delivering 828 × 57 mm rockets.
- ATGM Chetak: 3 Sorties delivering 4 × AS-11 missiles.

Night Strikes During this period, it was also decided to carry out night strikes by Mi-25s under artificial illumination to target LTTE boats moving along the coast at night. Eight sorties were flown by Canberras of 6 Sqn dropping 11 flares between 2nd to 19th Aug. Missions were called off due to identification problems of LTTE boats among hundreds of fishing boats (despite total embargo on night fishing). Four sorties of Jaguars were also launched over the Checkmate areas, dropping flares on the East coast for psychological pressure on the LTTE holed in these camps with IPKF troops tightening the cordon around them.

Air Strikes (23rd Aug 1988) Prior to own troops moving in for the final sweep of the area, heavy air and arty fire were carried out. At this stage, the troops earmarked for the sweep were already very close to the target area. In fact only about one square km was available within this cordon, which could be engaged from the air without any danger to own troops. Since the featureless jungle terrain precluded accurate navigation and target acquisition form the air, an AOP cheetah acted as an airborne FAC to get indicators in addition to the smoke indication by forward localities. Because of the operational problems envisaged, detailed planning/briefing was arranged for all participants on airborne recce/ground smoke indications, and operations were closely supervised.

Mi-25 The helicopter fitted with Doppler navigation system overflew a known pinpoint close to the target and flew on the given course and distance/time before releasing its bombs. Smoke indications were provided at the four corners of the cordon. The airborne FAC was in contact with the ground posts and the attacking Mi-25. Though the bombs were dropped within the target area, as confirmed by the airborne FAC, ground reports received from forward troops (1/1 GR) stated that one of the bombs exploded within 300 metres of them and a few shrapnel flew towards them cutting through their camouflage covering. It was later confirmed that the Gorkhas had moved ahead of their allotted position. The bombing missions, nevertheless were called off after the sortie because of safety considerations of own troops. It was too much to presume that each ground post would be exactly where it was supposed to be (due to the similar problems of accurate navigation/ground position for the ground forces as well as the aircrew in the thick jungle terrain).

Mi-8 The crew did a recce of the target area in an Army Cheetah, before proceeding in the lead Mi-8 for the strike. With the help of an airborne FAC and ground smoke signals, the Mi-8s, led by DN Sahay, CO 119 HU successfully carried out seven live missions that day, pumping in 430 rockets precisely within the target area. In order to save the turnaround time, two additional loads of rockets were brought in from Trincomalee and offloaded along with technicians to Mullaitivu. After the first live missions, consisting of a pair of Mi-8s, the two helicopters landed at Mullaitivu, rearmed and were airborne for their next mission within 45 minutes. The whole turn around and loading was done in field conditions without any base support whatsoever.

Indian troops confirmed the presence of a cache of arms/ammo on top of a tree, supported on a huge machan like structure. An AOP Cheetah fired a few rounds to indicate the target in four dives. However all four missiles fell short. No further mission could be mounted as ground combing had been ordered. On 23rd Aug, the largest ever induction of troops (over 750) was carried out by Mi-8s in a single day. On 24th, though own troops had started closing in, three Mi-8 missions were launched at dawn and a total of 192 rockets were fired, all landing precisely in the target square.

LTTE Attack on TACT HQ In a desperate attempt to get out of the cordon and escape from further air attacks, militants carried out a last ditch attack on the TAC HQ at Nithikaikulum at 1400 hr on 25th Aug 1988. One group fired two RPGs, one at the parked Chetak there, scoring a direct hit and second at the command post tent where the GOC IPKF, AOIC, GOC 4 Div, Cdrs of 7 Bde and Special Forces were conferring. This RPG narrowly missed the tent. The militants then started firing their AK-47, but withdrew back into the jungle once own LMGs opened up. An armed Mi-8 at Mullaitivu, piloted by W/C DN Sahay, was overhead within 10 min and fired RPs in salvo in likely getaway direction as indicated by the orbiting AOP aircraft. Despite failing light and bad weather, Mi-8s successfully inducted troops to reinforce the Nithikaikilum garrison. The reaction of the Mi-8 to meet this special contingency both in the strike and induction roles was superb and highly appreciated by the ground formation commanders. On the same night, two more attacks were carried out by the LTTE. To prevent a breakout, additional troops were inducted by Mi-8s early next morning.

A Daring Cas Evac On 6th Nov 1988 around 2100 hrs, a distress call for emergency casevac was received. The meteorological conditions prevalent were light drizzle, poor visibility and strong winds. However, the met report was overruled in view of the experience, confidence and determination of the detachment crew of 119 HU. The helipad at Nuttur was lit with headlight of four vehicles. The Mi-8 landed with precision and skill and safely picked up the casualty and an EPRLF supporter, who was seriously wounded. All throughout, direct casevac by An-32s from the Island airfields to rear airfields in the mainland viz, Bangalore, Pune, Secunderabad were also carried out. Similarly, casualties were transferred from Madras rearward as and when needed, to make room for new/more serious casualties. Casevac by AN-32s and Mi-8s was given due priority and close attention.

Decision to Hold Elections The most momentous decision taken during this period, was the go ahead given for the provincial election. On instructions from New Delhi, IPKF declared unilateral cease fire on 15th Sep 1988 for five days to give LTTE an opportunity to hold consultations, lay down arms and join the democratic process. On 17th Sep, Mi-8s dropped propaganda leaflets over Jaffna town as per requirement of the Special Action Group (SAG). The cease fire was extended by another five days up to 25th Sep. There were no incidents of violation of cease fire by the LTTE. However, in retrospect it appeared that LTTE had used this opportunity to regroup and re-equip for renewed violence as there was no response from LTTE for any peace negotiation. Immediately after the ceasefire, 'Op Chakravuya' was launched on 25th Sep.

Towards the end of Sep 1988 (22nd to 30th), night air maintenance by AN-32 to Jaffna and Trinco was carried out on a trial basis in order to ensure that our pilots as also the ground organisations would be prepared should such a requirement come up. On 9th Oct, as a gesture of political goodwill, 103 Sri Lankan prisoners were released from Boosa Camp in South-West Sri Lanka and brought to KKS Port in Jaffna in an Indian Navy ship. Thereafter, two An-32s airlifted 90 of them to Trincomalee, Batticaloa and Colombo as per their choice. One Mi-8 sortie was utilised to transport 12 prisoners to different locations in the Jaffna peninsula. As a prelude to holding provincial election and as on overture to the LTTE, their political leader, Krishna Kumar alias Kittu along with 137 LTTE inmates were released from various jails in Madras and Madurai and were flown to Jaffna and Trincomalee respectively in a specially planned pre-dawn operation. The prisoners were released simultaneously from their respective jails at midnight on 8th and 9th. As per the air plan, four AN-32s took off from Tambaram for Madurai at 10 minute intervals, the first being airborne at 0300 hrs. Three aircraft then took off at 0500 hrs from Madurai for Trincomalee with their special load as planned. All four aircraft took off for Jaffna between 1230 and 1300 hrs from where the prisoners were further airlifted to KKS, Jaffna town, Pt Pedro and Mannar in six Mi-8 sorties. Kittu was released by the IPKF town commandant in Jaffna town.

The Pre Election Environment The announcement of elections meant a sudden flurry of political activities involving movements of various political representatives, Indian High Commission officials, press and other personnel. There were innumerable demands for airlift, for moves between the mainland and the Island and also for the movements within the Island. The ubiquitous demands for air maintenance and transportation continued unabated. The utilisation of Mi-8s was highest during this period. The frequency of the SHBO missions by Mi-8s escorted by Mi-25s was increased as election approached to keep pressure on LTTE and prevent any large scale violence. The final go ahead to hold the provincial election was given in mid October 1988 and the date finalised for 19th Nov. To ensure that the IAF fulfils its commitments fully and effectively, the TAC HQ of IPKF was activated at Trincomalee. Eight AN-32s were positioned at Colombo (Ratmalana) at 0730 hrs on 17th Nov for the airlift of election officials to Trincomalee, Batticaloa and Amparia.

The Elections The entire air movement plan was formulated by TAC HQ IPKF which involved airlift of officials to and from Colombo, their distribution to various polling stations along with ballot boxes, movement and recovery of officials and distributing ballot boxes to selected points. In a few cases, the ballot boxes had to be retrieved by night. The last of the ballot boxes was safely picked up from Muttur at 2100 hrs on 19th Nov by one of our Mi-8s and delivered to the concerned officials at Trincomalee. Next day, the election officials were flown out to Colombo in an AN-32, complemented by SLAF YAK-8/12. The detailed planning ensured that the election process went through smoothly. The provincial election was followed by the Presidential Election on 19th Dec 1988 and the Parliamentary Elections on 15th Feb 1989.

The contribution of the IAF in ensuring that the IPKF fulfilled its mandate for Phase II i.e. to create a suitable security environment, for the holding of provincial elections had been truly immense. Nothing could operate or function during this phase including holding of elections without air support, and the IAF provided that support unfailingly and unflinchingly. The post election mandate of IPKF was to support and assist the newly elected Provincial government. The role of the IAF during this Phase III of the IPKF action in Sri Lanka was as crucial as in Phase I and Phase II.

Phase III of IPKF Operations in Sri Lanka (Feb to Apr 1989) In the post election scenario, redeployment of own troops had to occur both from the point of view of countering LTTE operations in the Vavuniya sector and increasing the effectiveness of the newly raised CVF (Civilian Volunteer Force) as part of the SLP (Sri Lankan Police) in the North Eastern Province. These redeployments took place entirely by road except in certain areas of Vavuniya sector (Alampil jungle area). Air effort for these redeployments amounted to 28 sorties /5.45 hours, between 18th and 22nd Feb. On 2nd March, two companies of 6/8 GR made contact with militants in the area west of the Nayaru Lagoon and south of Kumulamunai. Fire was exchanged till nightfall and recommenced throughout 3rd Mar. 566 troops and 56.66 tonnes of load were inducted into the area between 5th and 9th Mar. The air effort involved was 209 sorties amounting to 29 hours. The tasks included induction, de-induction, re-deployment and casevac.

In the same period, speculative fire against identified target areas by Mi-25s amounted to 16 sorties in 11.20 hours, expending 20×250 Kg Bombs, 600×57 mm rockets and 900×12.7 mm front gun rounds. Between 3rd and 13th Mar, for the first time Mi-8s were used for bombing missions; seven sorties in six hours were flown on the Mi-8 expending 111 rockets and 12×250 kg bombs. After nearly four to five days of fighting, the LTTE had lost some 49 killed and a large number injured. It was estimated that the militant's strength in the area was 250-300 in several camps. When 6/8 GR had come in contact with the militants in a thick jungle about a kilometre west on the top end of Nayaru Lagoon, an Army Cheetah tried to land in a hastily prepared clearing in the jungle (the aim was to get a briefing on enemy dispositions and bring in some supplies). When the aircraft was in the last stages of approach 'something whizzed past the aircraft in front'. The captain initiated an immediate take off and recovered the Cheetah safely. The projectile may have been an RPG or even a SAM.

Op Shamsher In view of the situation in the Alampil jungle area, Op Shamsher was conducted between 18th and 30th Mar. It started with induction, re-deployment and finally de-induction of troops. Op Shamsher resulted in the destruction of all major camps of the LTTE in the jungle areas

west of the Nayaru Lagoon. LTTE transmission intercepts indicated that
this operation had a telling effect on them, both in terms of losses in
manpower and arms/ammunition. Intercepts also revealed that the LTTE
had suffered a loss of 125 killed, 59 injured and two surrendered. The total
effort flown including armed missions was:

• SHBOs for induction and de-induction :	262 Sorties/74:05 hours
• Re-supply and Casevac (MI-8s) :	154.073 tonnes of load and 1276 troops.
• Armed Mission (Mi-25s) : (702×57 mm rockets and 56×250 kg bombs)	34 sorties/25:00 hours
• Armed missions (Mi-8s) : (127×57 mm rockets)	Three sorties/01:15 hours

General Trend of Air Operations Operational missions both on the Mi-8
and Mi-25 were concentrated mainly in the Vavuniya sector, where militants
had established themselves in a belt running from Mullaitivu on the eastern
coast, through Oddusudan, Nedunkeni, Omantai/Puliyankulam on the
Kandy road and eastwards through the jungles northwest of Vavunia to
Adaman/Mannar on the west coast. Operational missions in the other
sectors were minimal. A large part of the air effort utilised was towards
meeting the IPKF mandate of assisting the newly elected provincial council
in establishing themselves and developing effective control over their area
of responsibility. There was of course the all important air maintenance
without which own troops would have found it indeed difficult to function
effectively. In fact, air maintenance for fresh rations, mail and casevac were
given very high priority. Wherever possible, air maintenance was avoided
such as in the Jaffna peninsula, where non use of air effort served two
purposes, firstly, reduction in the overall air effort, and secondly, the more
important factor of being seen to have the capability to maintain own troops
by road despite militant interference.

April 1989 ended with two incidents wherein air operations played a
major role. One occurred in the area immediately to the north west of
Trincomalee where a Mi-25 with 128 rockets was used. The targets were
LTTE hideouts. The General Officer Commanding 36 Infantry Div Maj Gen
Jameel Mahmood, an Air OP officer was himself on board the Mi-25 to
witness on 13th Apr and 16th Apr. On acquiring the target, 127 rockets were
fired. A LTTE transmission intercept confirmed that three militants had been
killed or injured, apart from the hideout being damaged badly. The second
operation occurred in the Vavunikulam area south east of Killinochi.

Information regarding large training camps in these jungles had been received over a period of time. A ground operation with troops and air/ artillery support was planned. A large number of troops were inducted into the area by air on 21st Apr. For this SHBO, one Mi-8 and one Mi-17 flew 59 sorties in 16.10 hours. 628 troops were inducted and redeployed. At this stage, a decision was taken to carry out this operation without troops. Some of the troops were moved back to their original locations and the rest were used in stop positions around the target areas. The whole area was then selectively bombarded by artillery and Mi-25. On 23rd and 24th Apr at selected areas in the jungles where hard intelligence on exact locations of camps was available, the Mi-25s flew 13 sorties in 13.35 hours and Mi-8s flew four sorties in 4:25 hours. In all forty 250 bombs and 122 rockets were expended.

May–June 1989 This was the period when the LTTE overtures to the SLG had culminated in the opening round of talks between the two in the second week of May. The major event involving air operations was deliberate operations against LTTE in the Nedunkeni area. It was confirmed that more than 21 LTTE cadres were killed in this operation. In one incident when own troops were ambushed near Puliyankulam, an airborne Ranjit was ordered to the area. The Ranjit engaged the LTTE with machine gun fire. In this process one militant bullet hit the ac. The bullet penetrated the bubble, hit the co-pilot's hand on the control column, ricocheted and hit the captain in the jaw. The captain, Maj D Sahgal and his co-pilot Capt Karve, though injured, managed to recover safely at Vavunia. Poor tactics and the vulnerability of the Ranjit were evident in this incident.

Apart from such major events as above, air operations during May and June 1989 centred on minor troop movement of provincial Govt (PG) members. The Mi-25 was used mainly in the role of speculative fire and bombing of known LTTE hideouts in the jungle. This effort was correlated with intelligence information, both from radio transmission intercepts and ground source reports. These reports definitely indicated that such air effort with well placed and planned patrols curbed the operational capabilities of the militants. Their movements were considerably affected and their camps were being hit continuously. Apart from this, one specific mission with two Mi-25s armed with 4 × 250 Kg bombs and 128 × 57 mm rockets over the Mawilla jungles south of Trincomalee is worth mentioning. This area was assessed to have a large number of LTTE camps with strength of approximately 150 militants. The sortie was planned and executed around 1500 hrs on 28th June. Subsequent intercepts revealed that 25 militants were killed and 35 injured directly as a result of this mission. It was assessed that one of the bombs had hit an ammo dump in the area. The total effort flown during these two months was as follows:

- Mi-25 armed missions – 95 sorties/76:00 hours expending 1494 × 57 mm rockets, 146 × 250 kg bombs and 860 × 12.7 mm front gun ammo.
- Mi-8 armed missions – 35 sorties/28:20 hours expending 1039 × 57 mm rockets.
- Mi-8 SHBOs – 170 sorties in 75:40 hours achieving 28.402 tonnes load /1257 troops.
- Mi-17 SHBOs – 86 sorties/39:20 hours, lifting 71.11 tonnes of load and 571 troops.
- Special Action Group – 136 sorties/60:30 hours using Mi-8 and Mi-17.

Jul-Aug 1989 In addition to effort utilised for air maintenance and assistance to the Provincial Government (PG), the following was flown towards pure military operations and casevac:

• Mi-8 SHBO	55 sorties/14:05 hours/9.1 tonnes of load and 327 troops.
• Mi-17 SHBO	193 sorties/65:20 hours 120.732 tonnes of load/1538 troops.
• Mi-8 Armed missions	10 sorties/3:10 hours/192 × 57 mm rockets.
• Mi-17 Armed missions	10 sorties/3:10 hours/192 × rockets.
• Mi-25 Armed missions	129 sorties/87:50 hours expending 240 × 250 Kg bombs 114 × 57 mm rockets and 215 × 12.7 mm F/G rounds. These were mainly distributed over the Alampil jungles west of Nayaru lagoon, the Andaman area South East of Munnar, the Mawilla jungles South of Trinco and militant pockets South of Batticloa.
• Night Casevac Missions	15 sorties /5:55 hours evacuating 31 casualties, some of them between 2200 hrs and midnight.

Night Bombing It was known that militants moved only at night for logistics purposes. Till around mid Jul 1989, some armed missions had been mounted around sunset time. It was now decided to launch a few missions at selected times during the night and early morning. The Mawilla jungles and Nayaru Lagoon area were chosen for this purpose. On 29th Jul, the AO planned and briefed the Mi-25 crew for night bombing missions over Mawilla jungles. The first mission with eight 250 kg bombs was carried out at 1930 hrs and the second one with eight bombs again at 2305 hrs. During the second mission, the crew sighted flickering lights and targeted their bombs on the area. Subsequent to these missions, effort was planned in the Nayaru Lagoon area also. In one of these missions lights were seen by the crew and treated as the target. This mission was at 0530 hrs on 28th July. Such night missions were continued for a short period during August also. However, it was never possible to obtain any information on the results achieved. In July, total effort towards night armament work was: Mi-25, 15 sorties in 11.05 hours expending 48 × 250 kg bombs; Mi-17 – one sortie/ 1.15 hours using 128 × 57 mm rockets.

To cater for various contingencies, as many as twenty Mi-8/Mi-17 helicopters had been planned for deployment at the four bases on the island, with a backup of five more. Dispersion had to be planned in terms of providing facilities at a large number of helipads to accommodate at least 20 × Mi-8/Mi-17, three Mi-25s and Army Aviation aircraft. These had to be such that all operational contingencies were met as they arose, at least for a period of 48 to 72 hours from the dispersed locations. Therefore, it meant dispersion of ATF, armament and personnel to ensure first line servicing and so on. During the month a number of contingency operations had been planned. These plans catered for an air support organisation from the mainland, establish a standard air support organisation and an Air Defence set up in the southern peninsular region of India. The announcement that governments of India and Sri Lanka had agreed to hold discussions on the subject of withdrawal of IPKF and allied matters came as an anti-climax, from the point of view of the forces on the ground. However, there was little let up in the activities and attitude of the LTTE in the North East Province or in that of the JVP in central and Southern Sri Lanka. Till 18th Sep 1989, when the joint communiqué was issued, there was no change in the IPKF activities or military policies. However, there was an air of expectancy, uncertainty and doubts about what were likely to be the future courses of action.

Air Aspects (Mar-Sep 1989)

- Continued heavy bombing of known areas of militant concentrations. Much damage in terms of physical losses and prestige was caused to the LTTE due to the fact that more authentic information was available through interrogation of captured militants.
- HQ Southern Air Command halted heavy bombing on 8th Oct 1989 and

limited the use of 250 Kg bombs. The AOC-in-C permitted bombing missions to the extent required essentially, but restricted the number of bombs in a mission to two. The Mi-17 helicopter was used for the bombing role for the first time.

- The TAC HQ at Trincomalee was wound up on 31st July 1989 but was re-established on 1st Aug. On 17th August, the AOIC returned to Madras and SASO Air Force cell was moved to Trincomalee maintain the TACT HQ.
- Induction/de-induction of troops and armed missions in connection with two deliberate operations, one in the Alampil – Nedunkeni area and the other in area Norht West of Trincomalee. 100 kg bombs were inducted for use instead of 250 kg bombs, during the first week of September. Air maintenance by AN-32 and Mi-8/ Mi-17 continued.
- Total effort flown for the deliberate operations in Alampil and Nedunkeni area, stretched over the period 1st-16th Sep 1989. The total effort towards induction, re-deployment, casevac, resupply of water, cooked food and other items and finally de-induction was:-
  - Mi-8 SHBOs   -57 sorties/17.30 hours /3.775 tons and 484 troops.
  - Mi-17 SHBOs -181 sorties/39.55 hours /95.612 tons of load and 1046 troops.
  - Mi-25   -9 sorties/8.10 hours / 7 × 250 Kg bombs and 142 × 57 mm rockets.
  - Utilisation Rates: An-32 -3.26 tonnes/hours; Mi-8 -2.5 t/hr; Mi-17 -3.06 t /hr.

Oct 1989 – Mar 1990 The joint communiqué of 18th Sep 1989 called for the suspension of offensive military operation by the IPKF. From this date onwards therefore, all air effort was towards air maintenance, SAG tasks and casevac only. Use of air for de-induction was essentially for light/ delicate equipment bought out of contingency funds that had to be moved to Trivandrum and armament that had to be back loaded to Sulur. The sorties planned for this requirement were mostly those returning after routine air maintenance tasks to ensure economy of air effort. Special air effort of IL-76 was planned for the T-72 tanks only from the Jaffna sector.

Worst-Case Scenario Towards the last phase of de-induction of IPKF, intelligence indicated that the LTTE, taking advantage of reduced strength, might indulge in attacks on the Indian Forces. This was likely from 1st March 1990 onwards when own forces would have thinned out with only minimal fire support being available. In consultation with HQ SAC and Air HQ, the following deployment and employment was executedbetween Mar-Sep 1989:

Date	Operation	No of ac utilised	Load/Troops carried
03 Mar 89 to 21 Mar 89	Op Campus Armed Task at Mullaitiva, Alampil, Nayaru Lagoon area.	2 × Mi-5 (27 sorties)	Arm expended. R/Ps 855 F/G 900 250 Kg
23 Mar 89 to 31 Mar 89	Op Shamsher Armed Task at Mullaitivu west of Nararu Lagoon.	2 × Mi-25 (24 sorties)	Arms expended R/Ps 447. Bombs 250 Kg 38.
20 Mar 89	Op Campus SHBO Induction of 19 Mahar, 1 Para, 9 Para, Engrs, 14 Mahar, 51 Engrs.	5 × Mi-8 (93 sorties)	27.750 Ks/557 troops
23 Mar 89 to 24 Mar 89	Op Shamser SHBO Induction & re-supply at Mullaitivu, Alampil, Nayaru Lagoon.	5 × Mi-8 (85 sorties)	31.420 Kg/384 troops
23 Apr 89 to 24 Apr 89	Armed Task Mi-25 SE of Killinochi	2 × Mi-25 (15 sorties)	Armt expended 250 Kg Kg Bombs 40
21 Apr 89	SHBO Mi-17 at Jaffana Sector- Induction of 16 Raj Rif, 7 Bihar, 5 Para.	2 × Mi17 (27 sorties)	325 troops
17 Mar 89 to 18 May 89	Armed Task Mi-25 in Jaffna	1 × Mi-25 (8 sorties)	Armt. expended. R/Ps 298, 250 Kg bombs 18
14 Jun 89	SHBO Delft Inland Ops – Induction of 18 Garhwal	4 × Mi-8 (25 Sorties)	241 troops
14 Jun 89	Mi-25 Armed Task-Escort for Mi-8 induction	1 × Mi-25 (2 sorties)	Nil Arms expended
16 Jul 89 to 17 Jul 89	Airlifting of Late Amrithalingam's body, PC Members and Mr. Natwar Singh, Minister of State External Affairs.	3 × Mi-8 (24 sorties)	Total of 82 personnel
19 Jul 89	SHBO Mi-17 VVY Sector Induction of 7 Para, 9 Para, 10 Para & 1/5 GR	4 × Mi-17 (33 sorties)	30,140 Kg/356 troops.
29 Jul 89 to 31 Jul 89	Armed task Mi-17 in VVY sector (First time Mi-17 utilised for Armed Task role)	1 × Mi-17 (33 sorties)	Arms expended. RPs 312

(Contd.)

Date	Operation	No of ac utilised	Load/Troops carried
01 Aug 89 to 03 Aug 89	Armed Task Mi-17 in VVY sector Day/night bombing sorties.	2 × Mi-17 (10 sorties)	Arms expended 250 Kg bomb 36.
04 Aug 89	SHBO De-induction of 4 Para in VVY sector.	3 × Mi-17 (14 sorties)	18,100 Kg/304 troops.
01 Aug 89 to 08 Aug 89	Armed Task by Mi-25 in VVY sector	3 × Mi-25 (23 sorties)	Arms expended.
12 Aug 89	Mi-25 Armed task escort for Mi-17 CAS EVAC Adampan	1 × Mi-25 (1 sorties)	No Arms expended.
13 Aug 89	Mi-25 Armed Task, escort of convoy	2 × Mi-25 (3 sorties)	Arms expended 250 Kg bombs 06 F/G Rds.
02 Sep 89	SHBO induction in Alampil area of VVY Sector.	2 × Mi-17 (15 sorties)	11,200 Kg/102 troops
05 Sep 89	SHBO Tactical Induction VVY sector. Alampil area and re-supply of water to troops	3 × Mi-17	19,005 Kg/427 troops
05 Sep 89	SHBO. Tactical induction in Alampil area of VVY sector	2 × Mi-8 (37 sorties)	12,735 Kg/286 troops
12 Sep 89	SHBO de-induction	2 × Mi-17 (19 sorties)	37,750 Kg/126 troops.
13 Sep 89	Armed escort of convoy Jaffna	1 × Mi25 (04 sorties)	No arms expended.
14 Sep 89	Armed task Mi-25 Alampil area	2 × Mi-25 (19 sorties)	Arms expended 100 Kg bombs 10, R/P 668.
18 Sep 89	SHBO induction and de-induction of troops in VVY.	4 × Mi-8 (28 sorties)	7290/261 troops 8000/120 troops

Two Gunships were maintained at Ramnad from 1st to 24th Mar 1990. A flight each of Jaguars and Mirage-2000 operated ex-Bangalore and Trivandrum respectively from 1st to 24th Mar. Three IL-76s were deployed at Madras airport for re-induction of reserves and armament, if necessary. An option to call for four additional AN-32s from Training Command was kept open. Additional Mi-8s were kept at Sulur to be ferried to the island, if necessary. Apart from the planning process outlined, the operational responsibility of the IPKF in areas still occupied were also catered for. In the Jaffna peninsula, this included joint patrolling of the Jaffna lagoon, on either side of Elephant Pass, by Army, Navy and Air Force. A few helicopter missions along with Army Engineer boat patrols were carried out. These were stopped when further de-induction from Jaffna peninsula was ordered. Also, during the period of mid-Dec 1989 to early-Jan 1990, aircraft transiting between Trincomalee and Jaffna had to stay about a kilometre out into the sea since flying over vacated areas was not permitted. Two Air Traffic Control posts were, therefore, established; the northern-most in the Trincomalee sector and the southern-most in the Jaffna sector to monitor traffic over the sea and off the vacated Mullaitivu area. With further de-induction being ordered, these ATC posts also had to be pulled back. Thus, from 15th Jan 1990 onwards aircraft transmitting between Trincomalee and Jaffna were instructed to fly as high as possible, with the aim of staying within radio contact with either Trincomalee or Jaffna.

Apart from the ongoing air maintenance task, the IAF had two major operational tasks during the process of de-induction. These were the move of 9 Para from Vavunia to Batticaloa on 20th Nov 1989 by Mi-17/Mi-8 and aerial protection of convoys between Batticaloa and Trincomalee by Mi-25 gunships. While the move of 9 Para was a routine operation from the point of view of the IAF, the convoy protection operation needed a fair amount of co-ordination and planning. Two army Ranjit helicopters, two armed Mi-8s and one Mi-25 were used for this task. The time taken for a convoy to cover the distance between Batticaloa and Trincomalee was approximately 10 hrs. This period was broken down into blocks of two hours each and each type of aircraft was allotted alternating blocks. At that time there were 800 odd prisoners held in various IPKF camps. A military decision was taken to move them to two central locations; one at Trincomalee and one at Jaffna. The major portion of transfer was done by LST on 14th Nov 1989 and only 10 militants were moved by Mi-8 on same day. From Vavunia, a total of 89 inmates were moved by Mi-8 to Jaffna in the first week of December. By Feb 1990, all were released in small batches.

Further Developments   On 20th Jan 1990, areas along the neck of the Jaffna peninsula had been fully vacated. The LTTE had progressively occupied the areas in force. On 21st Jan, LTTE resorted to unprovoked firing on an IPKF line party moving from Battalion HQ to Jaffna fort. This situation was contained within 72 hours. An armed Mi-8, a Mi-25 and a Ranjit were used. The targets given to the Mi-8 and Mi-25 were in the open and easily seen and destroyed. The third week of Jan 1990 saw some serious and major clashes between the LTTE and IPKF. A Chiefs of Staff Committee directive was issued, thereafter, in which two contingencies were planned for:

- Worst case scenario of the de-induction, calling for re-induction of reserves from on board Naval Ships and/or mainland to stabilise the situation and continue with de-induction as a fighting withdrawal. This became known as Op Pushpak.
- Re-induction of reserves and additional forces to from a Bridge Head in the Trinco sector and continue de-induction from the Jaffna sector. This would have meant a partial Op Round Up in the Trinco sector. This operation was called Op Garuda.

The minimum force of the IAF was a pair of Mi-25 and Mi-8 each at Trincomalee and Jaffna. The Mi-8s were with armament capability. On 20th Mar 1990, when Jaffna was to be vacated, two Mi-25s and two Mi-8s were to proceed to Ramnad and standby there for re-induction, if it became necessary. IAF assets were progressively de-inducted either by air by returning AN-32 flights or by road convoys to Trincomalee for onward de-induction by sea.

Battle Damages   Records of Battle Damages that were available are only from 15 Oct 87 onwards are reflected.

Date	AC Type	Brief Details
22 Oct 1987	Mi-8	AC was hit soon after getting airborne from a helipad 12 Km from Jaffna. AC landed at helipad and flown back to Jaffna after inspection.
03 Nov 1987	Mi-25	AC hit by Militant fire in area west of Palaly while carrying out recce for LTTE vehicles. Recovered safely.
03 Nov 1987	Mi-25	AC hit by Militants fire in area west to Palaly while carrying out recce for LTTE vehicles. Recovered safely.
15 Dec 1987	Mi-25	Bullet holes noticed after AC landed. Nothing untoward was noticed during the mission.
23 Jun 1988	Mi-8	AC was hit by ground fire while attempting to winch troops into a clearing in the jungles.
08 Aug 1988	Mi-8	Ground fire caused damage to rotor blades while flying low over militant positions.

Date	AC Type	Brief Details
07 Feb 1989	Mi-8	AC hit by a stray bullet while coming into land at Pt Pedro.
03 Apr 1989	Mi-25	After a bombing run AC was descended to assess damage. On landing bullets holes were seen on rotor blades.
16 May 1989	Mi-8	Accidental fire by a Jawan while boarding the AC caused damage to the starboard engine.
20 Jan 1990	Mi-8	While carrying out eighth pass over target area AC was hit by LTTE fire. AC was recovered safely.

Besides the above incidents, there were two cases at a place called Tunukai in the Vavunia sector, which have not been reflected as battle damage: A Mi-8 was a write off, when it toppled over just before touchdown because of a tree stump hidden behind bushes. It was carrying 2.4 tonnes of artillery ammunition, but no fire occurred and the ammunition was recovered. In the other case, while troops were entering a Mi-8 with its engines running, one Jawan walked into the tail rotor. The Jawan was killed. Minor damage sustained by the helicopter was repaired.

Mi-17 Flying Effort (Jul 1987 to Mar 1990)

Prior to commencement of Op Pawan and up to 11th Aug 1987, a Mi-17 Detachment (129 HU) operated in Sri Lanka which was withdrawn subsequent to the move of 109 HU (Mi-8s) to Sulur. The detachment achieved 75 sorties in 39 hrs of flying. It lifted 25 tonnes load and 591 troops in various roles. When a large number of Mi-8 engines were withdrawn and the serviceability state became low, Air HQ decided to revive the Mi-17 detachment.

Task	Sorties	Hours	Load (Tonnes)	Troops/Pass.	Total Load (Tonnes)
SHBOs	606	209:00	392:590	5963	988.89
Armed Mission	26	09:40	R/Ps-312		
Bombs	Nil	N/A			
Casevac	53	17:40	Nil	116	11.6
Air Maint	3366	711:45	791:809	12180	2009.809
Comn	479	243:05	171:158	4480	619.258
SAG Task	354	183:00	66:313	2645	330.813
Other Op Tasks	87	53:35	66:044	355	101.544
Total	4961	1428:35	1487:914	25740	4096.192

Mi-8 Flying Effort (Jul 1987 to Mar 1990)

Jul 1987–Feb 1988: Detailed record is not available. Consolidated flying effort carried out is as follows:

Sorties	Hours	Load (Tonnes)	Troops/Pass.
6068	1861:00	2183	Not avlbl

Mar 1988 to Mar 1990

Task	Sorties	Hours	Load (Tonnes)	Troops/Pass.
SHBOs	5821	2265:10	1208.705	49642
Armed Mission	153	129:40	R/Ps-4814	
Bombs	Nil			
Casevac	3398	144:15	6.2	865
Air Maint	2419	2978:05	3216.858	3839
Comn	972	559:35	147.298	5888
SAG Task	250	183:10	57.313	1333
Other Op Tasks	463	850:25	161.207	7012

The Gallant Knights-109 HU (From the Unit Diary)

In the first week of June 1987, six An-32 aircraft of the IAF escorted by the Mirage 2000 quietly took off from Bangalore and in a surprise move air dropped food supplies and medicines over Jaffna. Amidst the noise of the media and the flurry of political statements the IAF began to quietly prepare itself. The writing on the wall was clear – the involvement of Indian forces in the issue was inevitable. The famous accord of Jul 1987 between the Indian PM, Rajiv Gandhi and Sri Lankan President, Jayawardane, followed on 29th Jul. Subsequently on 1st Aug, three Mi-17s of IAF landed at Jaffna. A week later, six Mi-8s replaced the Mi-17s. The old war horse was back in the game.

Initially, the role of the Mi-8 was restricted to only that of communication. Representatives of the various militant groups, mainly the LTTE, were carried to and fro between Batticaloa, Vavunia, Trincomalee and Jaffna. Troops of the IPKF were also deployed rapidly at various places by the Mi-8s. The helicopter, with its troop carrying capability of 24 fully equipped soldiers, was ideally suited for the role. After the signing of the accord and the subsequent induction of the IPKF it appeared as though the job for the IPKF would be easy. Talks were on between the Sri Lankan Government, the Indian High Commissioner and LTTE Chief, for setting up of an Interim Administrative Council. The Tamil militant detainees were being released by the hundreds from Sri Lankan prisons and the Mi-8s from Jodhpur, Yelahanka and Sulur were kept busy round the clock in transporting the freed prisoners to their destinations in the North Eastern Province of Sri Lanka.

All Tamil militants were granted general amnesty by the President of Sri Lanka. The various Tamil militant groups were surrendering their arms to the Peace Keeping Force. The LTTE apprehensively started responding to the efforts of the IPKF. Towards the middle of September it appeared as though at last peace would come to stay at the picturesque island of Sri Lanka, and it would only be a matter of time before the IPKF could sew up things nicely, wind up and head for home. However, the events of the next few days proved that appearances could be deceptive and that there was much more by the IPKF to be done, many more sieges to be laid, many more battles to be fought, many more lives to be lost before the IPKF could be done with it. During the first week of October, seventeen LTTE men were apprehended by the Sri Lankan Navy for allegedly smuggling arms into Sri Lanka by small boats. The prisoners were brought to Jaffna airfield to be kept in custody pending further transportation to Colombo. The Seventeen included two Area Commanders– Pullendzan and Kumarappa. The Indians were un-successful in their attempts to secure their release through negotiations with the Sri Lankan Government. When the moment for transporting the prisoners came, all seventeen committed suicide by biting on their cyanide capsules. The LTTE unleashed a reign of terror and violence leaving hundreds of bullet riddled burnt and mutilated bodies of innocent civilians in its wake. The LTTE had sealed its future and spelt out its own doom; and the IPKF was forced to take up arms against the LTTE.

The Indian Army faced the rather stiff task of freeing Jaffna from the clutches of the LTTE. The advance commenced on five different axes. It was then that one of the biggest helicopter supported operations in the history of the sub-continent was launched. On the night of 10th Oct 1987, three Mi-8s were involved in positioning a large number of troops and ammunition on helipads at Navatkuli, Mandai Tivu and Karai Nagar around Jaffna town. There was no incident of firing at the helicopters that night; however this happy state was not to continue for very long. Forces were being concentrated in the Jaffna Peninsula, and towards this the Mi-8s toiled all day on 11th Oct to position troops from Elephant Pass to Navatkuli. The supreme test for the tough Mi-8s and their gallant aircrew was yet to come. The orders were out for the LTTE cadres to shoot down all helicopters.

On the night of 11th October, a massive Special Heliborne Operation (SHBO) was planned. The aim of the mission was to induct 120 para commandos and about 400 troops of a Sikh Light Infantry battalion with arms and ammunition right into the stronghold of the LTTE in the Jaffna

University area. The task force consisted of a mixed complement of one aircraft from Jodhpur, one aircraft from Yelahanka and two from the Knights at Sulur, with Wg Cdr VKN Sapre as the Task Force Commander. The individual captains briefed their crew prior to take off, to expect small arms fire from ground. The co-pilots were briefed to take over controls of the aircraft in case the captain was injured. The aim was very clear – the mission must go through.

The aircraft got airborne on time at 0100 hrs, each with twenty troops on board. The weather was bad with clouds down to 200 meters, the moon totally obscured by the clouds, the night practically pitch dark. The black silhouettes of the four machines could be seen pressing relentlessly on towards the target. As the first wave approached the target and commenced its descent and deceleration, the militants in the vicinity opened fire from the ground. The firing concentrated on the landing helicopters from all the directions, however at that critical juncture, the ground fire was the least of the pilots concern for they were busy peering desperately out into the darkness looking for obstructions on approach. The helicopters were landing without any lights and the encountering of any obstruction would mean a swift end of the aircraft and souls on board.

After the first wave the LTTE managed to gather around the landing zone and occupied a multi-storeyed building on the approach path of the helicopters and the hail of lead became denser with even 0.5 inch calibre machine guns being used. In spite of the increasing battle damage to the helicopters, wave after wave of the Mi-8s kept coming in and taking off after off-loading the troops, until finally all the four helicopters were so badly shot up that the engines main rotors and tail rotors were critically damaged and the machines could not be flown any further. Some of the mechanical linkages too were severed and what kept the men and machines ticking, long enough to induct 150 troops and one tonne of ammunition is nothing short of a miracle. The heliborne operations for the night had to be called off and a road column was subsequently dispatched to marry-up with the heli-landed troops. This particular night operation brought the gallant aircrew four Vir Chakras and four Vayu Sena Medals (gallantry).

Early next morning, reinforcements for the battered fleet were flown in and operations commenced afresh. Due to the inaccessibility of the forward positions by road, the Mi-8s became the lifeline for the fighting troops in the Jaffna Peninsula. More than eighty sorties a day were flown from dawn to dusk in order to support the ever increasing members of the Indian troops. At Jaffna airfield streams of fixed wing aircraft were pouring in, inducting fresh Brigades and the scene looked busier than a busy International Airport at its peak hours. The further rapid deployment of these troops into the forward positions was left to the Mi-8s.

For the next ten days, the Mi-8s put their noses to the grind and just carried on flying. Large amounts of ammunition and rations were flown into forward positions, to helipads which were secure and at times to helipads which were not quite secure. The demand on the available Mi-8s in the Jaffna Peninsula was ever increasing. Under this intense pressure the Mi-8s rose to the occasion. Casualties were evacuated from areas where active engagement with LTTE was on. Artillery guns were flown from Jaffna airfield to the Jaffna Port. In response to SOS messages, small arms ammunition was air dropped in places where it was not possible to land. More aircraft were damaged due to small arms fire and yet the fleet did not falter. Repair work carried on side by side and the repaired aircraft would get airborne again and yet again. Special heliborne missions of inducting troops continued to be launched. The dauntless aircrew ensured that the lifeline to their brethren on ground never got interrupted. The handful of Mi-8s continued to achieve the impossible, proving that any task can be achieved by skill, grit and determination.

Finally on 23rd Oct 1987, Jaffna fell to the advancing Indian troops and was declared liberated. The Mi-25 gunships had moved into Jaffna Peninsula and tasted success when they were able to kill a large number of LTTE militants in an attack mounted on Chavakacheri. The Mi-25 in subsequent operations came to be dreaded by the LTTE. Having succeeded in evicting the LTTE from their strong holds in and around Jaffna it became the task of the Indian troops to rehabilitate the town and its citizens and to keep the town free of guerrilla attacks and the roads clear of ambushes. The LTTE did flee from Jaffna but left a large amount of arms and ammunition cached in places known only to them in the hope that they would one day regroup and recapture Jaffna. The months of Nov, Dec and Jan were spent by the Indian Army in flushing out operations and it became a routine for the Mi-8s to undertake special heliborne missions in addition to the normal logistics support sorties. SHBO missions were carried out at Kayates Island and in areas North-East of Vavunia in the dense jungles as the militants had taken refuge there. SHBO missions were also carried out.

By the end of February 1988, all Mi-8s from other units were withdrawn and the responsibility came to rest with the Knights (109 HU) who had already been operating since the beginning and the Stallions (119 HU) who had just then been shifted from North East. The two units dug themselves in and settled down to a commitment which was predictably going to be protracted and a long drawn out task. In the month of Aug 1988 the Indian

Army launched 'Operation Checkmate' in the dense jungles, North East of Vavunia for by that time; the militants had concentrated in this area and established training camps. Not unlike the month of Oct 1987, the month of Aug 1988 also saw the Mi-8s busier than bees. The tremendous logistical problems of supporting a large force in the midst of thick jungles for an extended period of two months could only be solved by the prowess of the mighty Mi-8s and the Knights once again levelled their lances. Landings were carried out in the jungles in unimaginable places for the express purpose of reaching cooked food and desperately required water barrels to the fighting troops. Casualties, with limbs blown off due to booby traps and land mines, which could not even be transported to the nearest clearing in the jungle, were brought out by winching up through the tall trees. Air to ground rocket firing was carried out by the Mi-8s with pin point accuracy. Time and again casualty evacuation was carried out even at night. The Mi-8s once again acquitted themselves well and saw the operation through.

Op Cactus Running true to form, the Knights did not confine themselves to one particular operation at a time. History stood testimony to the fact that Knights' desire to cross unknown frontiers and to press ever onwards was too strong to be suppressed. The desire manifested itself once again when on 29th Dec 88, they found themselves mounted on their battle steeds with their lances pointing southwards; charging to the rescue of the beautiful and picturesque Island of Male. For over three hours they flew over sea without the help of even a weather radar or any other navigational aid, which in itself was a record. On reaching Male they soon put things in order by inducting Indian Army troops into strategic positions and flying reconnaissance sorties. Having emerged champions of the tournament, the Knights returned to India on 3rd Jan 1989.

Rescue of an Aircraft The unit extended their missions of mercy not only to human casualties but also to crippled helicopters. In Jul 1989, when a Chetak helicopter was shot up by LTTE ground fire near Mettur in Trincomalee sector, the Army Aviation pilot was forced to make an emergency landing in an unsecured area. It was once again the Knights who levelled their lances and dashed to the rescue. Wg Cdr M Ramakrishna, the CO, personally undertook the mission. Mustering skills accumulated over years of flying, he set his mind to the task and successfully lifted out the Chetak, under slung to his Mi-8, to the security of Trincomalee airfield. The Knights had chalked up yet another victory against seemingly impossible odds.

119 HU in Op Pawan

The unit was moved to Sulur for Op Pawan in Sri Lanka on 23rd Feb 1988 and was re-christened as 'The Stallions'. It **flew 3800 hours** while undertaking multifarious roles ranging from SHBO, logistic support, communication, casevac etc. It air dropped **2820 tons of load and air lifted 42026** troops during Op Pawan. In recognition of the supreme sacrifices and service in keeping peace in Sri Lanka, the unit was presented a Tri-Services flag and was awarded one Vir Chakra and four Yudh Seva Medals.

The 125 Sqn Story: Diary Extracts

In the aftermath of the disastrous Jaffna University attack, the Indian Air Force realized the urgency of bringing in more firepower to support Indian Army operations and SHBO activities. The damage to the Mi-8s could have been avoided if there was more firepower to suppress ground fire and if the helicopters had additional armour to protect themselves. In Oct 1987, the Gladiators received a warning order for mobilising all available aircraft to Sri Lanka. Two aircrafts were immediately made ready and flown to Sulur. On arrival on 23rd Oct, Indian Army ordered us into battle. The first ever sea crossing was accomplished and we landed at Palaly. Immediately we were flown around in an Army Aviation Cheetah and familiarised with the area. The area around was largely held by the LTTE with their vehicles plying around with impunity. Meanwhile, an IL-76 arrived at Palaly with our logistics. On 25th Oct, we were given orders to launch an offensive with an aim to wrest the initiative from LTTE. The Gladiators immediately swung into action and struck a petrol pump and some moving vehicles on the road. The effect was immediate as the LTTE was now forced to abandon all movements by day and resort to only night moves. 11 Madras which was beleaguered between Jaffna and Elephant Pass was now able to get help from the garrison at Jaffna. Motivated by the success, IPKF immediately launched an offensive against the two urban bastions of Chavakacheri and Kodikamarn. The operation was a resounding success. By now, all formations wanted to have close support of the Gladiators. To meet these requirements a second detachment was started with one aircraft at Trincomalee.

We had compelled the LTTE to abandon Phase-I of their struggle i.e. frontal confrontation with IPKF, and led to initiation of guerrilla warfare from prepared camps deep inside the jungles. By now, the army was marching southwards restoring the vital rail road link between Jaffna – Vavunya. With that, all major urban centres along the route were safely in our hands.

Action at Mulai   On 3rd November 1987, 1 Para SF Commando was tasked to capture LTTE Targets at Mulai. The Gladiators dispatched two Mi-25s flown by Sqn Ldr Rajbir Singh and Flt Lt Atanu Guru to fly ahead and strike at militant strongholds. Rajbir arrived over the area and found himself in the midst of some heavy machine gun fire. He used his front gun to hit the ground targets in multiple runs. Two LTTE boats were destroyed in the initial attacks. While doing his third run flying against a Machine gun emplacement, he felt his Mi-25 shudder under hits from the ground. He immediately pulled out of the attack to assess the damage to this helicopter. He found his left engine RPM had fallen rapidly and was probably damaged. The Oil Pressure in the engine had also fallen rapidly indicating a heavy oil leak. Rajbir immediately switched off the port engine and commenced his return flight to base. He had a choice of jettisoning all his armament stores – the rocket pods under the stub wings which still had a few rockets in them, but chose not to do so as the stores would fall in rebel-captured area. Flying with the relatively heavy payload on a single engine would put enormous stress on the aircraft. A forced landing would mean that the aircraft and the aircrew will fall in the militant hands. For Rajbir, the choice was clear, he coaxed the aircraft back to Jaffna where he bought in the aircraft down to a smooth rolling landing. Subsequent inspection revealed that the R/T system was also knocked out by the ground fire.

Meanwhile a call was received by a beleaguered detachment of the Para Commandos who came under withering fire from the LTTE positions. Guru who was left holding the fort after Rajbir's departure assessed the situation.

The Para commandos were hardly three hundred meters away from the LTTE positions. There would be no room for error. As he came in to attack the ground positions, the LTTE militants directed a heavy stream of anti aircraft fire at him. Ignoring the streaks of tracer coming up to him, Guru continued firing his forward gun into the enemy positions. The militants' fire was no match for the battering they got from the Mi-25's weapons. They soon withdrew from their positions. Nearly a thousand rounds had been used up in this attack. The close air support from Mi-25s was so effective that 1 Para was able to overcome the LTTE positions. They managed to capture a flag belonging to the LTTE which was sent to the Squadron as a token of appreciation. A month later, on 7th December 1987, while flying in support of the para-commandos, the Mi-25 crew located a tractor transporting militants at Mulaitivu. The militants disembarked from the tractor and started firing at the Mi-25 helicopter. It returned fire accurately, destroying the vehicle and inflicting several casualties on the militants. This was later confirmed via radio intercepts.

A Captured LTTE Flag; Jointness Everywhere

A similar situation happened on 15th December, when a Mi-25 had to destroy a van carrying LTTE militants at Karaitivu. This time return fire hit one of the rotor blades damaging it. But the crew recovered the chopper back to base without further ado. These operations instilled a sense of fear among the LTTE against the 'Mudulai' which meant 'the Crocodile' as they had come to dub the Mi-25s. It was no surprise that a few months later, both Rajbir and Guru were awarded the Vir Chakra by the President of India. In its stay of two and a half years, from Oct 87 to March 90, the Squadron earned two VrCs and two YSMs. The much awaited withdrawal finally commenced in Jan 1990 with 57 Inf Div uprooting from Batticaloa followed by 4 Inf Div which vacated its position from Vavuniya. The det moved to Trincomalee. Jaffna was cleared on 19th Mar and the helicopters shifted to Rannad at the airfield. The final tactical withdrawal commenced on 23rd Mar with the last three aircrafts being flown back and the curtains being drawn on Op Pawan.

Through the Eyes of a Mi-8 Pilot
(Flying Officer NM Samuel)

On the morning of 14th Sep 1987, the aircrew of 110(Vanguards) Helicopter Unit, located at Kumbhigram, quite busy in their routine activities, were a bit puzzled to hear that two aircraft were to be dispatched to Sri Lanka to take part in the IAF operations there. The unit, positioned at this remote location for logistic support and SHBO operations in support of army units operating on the Bangladesh and Myanmar borders, consisted mostly of flight lieutenants and flying officers, with a couple of squadron leaders for supervision. As a flying officer with three years of service, I had a C/Green category on Mi-8, and, having been with the Vanguards for two years and fully ops for one year, was already considered one of the more experienced pilots.

But we had no idea, at that time, why helicopters had to ferry more than 3000 km from the north-east to Jaffna, or what we would be tasked for when we reached there. We reached Sri Lanka four days later, after halting for a day at Sulur. Not having flown over the sea before, the last part of the ferry was an enjoyable experience. It was a beautiful and picturesque country with long beaches and shallow waters on most of the coastal areas. As we approached the airfield of Palaly at Jaffna, we realised that, amid the seemingly serene environment, there was intense activity at the airfield. Except for the main runway the rest of the taxi ways and the dispersal areas were scattered with IAF aircraft. A couple of IL-76, a few AN-32s and quite a few Mi-8 helicopters were already parked in the limited area available.

We commenced operations soon enough, very early the next morning, after a short briefing. Procedures followed were quite simple – pick up vertically wherever the aircraft was parked, take off in any convenient direction on clearance from the ATCO, climb at the max possible power setting to be clear of small arms range, head directly to the area where troops were to be dropped, find a place to land, descend at the fastest safe rate, touchdown and hold with the aircraft light on wheels, get everything out as soon as possible, and reverse the whole routine to get back for another trip. We flew from sunrise to sunset, halting only for refuelling. We landed in water logged fields, on roads and mud tracks, on bunds between fields; we landed anywhere we could fit the wheels of the Mi-8 and in some places where we couldn't fit all the wheels, off loading from a low hover or with one wheel on the ground. We got used to the sound of gunfire and ignored it most of the time. We even learnt to refer to places like 'Puthukkudiyuruppu' and 'Periiyapuliyanklam' without getting tongue-tied in the process. The aim was clear – get the load off as soon as possible and go back for more. There was a lot of it coming into the island from India and the IPKF numbers were quickly growing.

Army troops were being flown in large numbers in transport aircraft, and most of them had very little idea of what they were there for. Some of them were quickly briefed on getting out of the transport aircraft and boarded onto helicopters immediately. Within hours they found themselves transported from their peace time locations and face-to-face with the Liberation Tigers in the Jaffna peninsula. The airfield was swarming with troops, vehicles and all sorts of ammunition and equipment. Aircraft were continuously arriving and departing in all directions. Somehow, in what seemed to be a complete chaos, the right equipment was loaded in the correct aircraft and delivered or inducted to the right place in a continuous process. How it was done, I never understood; perhaps a flight gunner will write an article some day! The operation soon spread to other parts of the northern part of the island and we found ourselves landing in other airfields – Trincomalee (China Bay), Vavuniya, Batticaloa, Mullaitivu. This of course only added to the uncertainty. One never knew where he would have to spend the night. We had to have a night kit in the aircraft at all times. After about a month of operating between all these airfields, I was asked to return

to my unit and I began the ferry back in Nov 87, thinking I had seen the last of the island. It was not be however, and I was back there before long.

I was posted to 109HU (Knights), Sulur in Feb 1988 and was back on the island within a few days of reporting to the unit. This time, I was there for much longer – on and off (more on than off) for almost two years. The Air Force detachments there were much better organised however, with an admin setup which was sent to each base for a period of three months. The IPKF had grown to a strength of over 80,000 and everyone realised that the detachments would have to be run for some time to come. The nature and quantum of air effort required continued and regular sorties were being flown both for troop induction and logistic support, as well as a few armament sorties when required. Most of the unit aircrew were based in Sri Lanka, going back to the mainland for short breaks. The accommodation and messing was better organised and the bases functioned with most facilities, including volleyball courts, in place.

A Joint Operations Information Room (JOIR) was set up in Madras, which controlled all Air Force operations on the island. Though we were still shuttling between air fields, I spent most of my stay at Vavuniya in the central part of northern Sri Lanka, which had a small airfield where An-32s could land. We also carried out a large number of missions with the Para-Commando battalions in Sri Lanka, and it was always a pleasure to be associated with any operation in which they took part. While all the troops that were inducted into the island were well trained, the three Para battalions, One, Nine, and Ten were definitely a class apart. We found them extremely professional in their approach, highly trained, understood the language of the aviators (much to our surprise!) and displayed excellent air sense. We constantly inducted them into areas where the Tigers were active and picked them up a few days later on completion of the operation. Heliborne operations with these troops were a memorable experience, as we would usually land in areas which were suspected of being hostile. The troops would disappear into the jungle within seconds of the wheels touching down and the aircraft could take off almost immediately.

Just Another Mission In the third week of Jun 1988, the JOIR received a request from the Army for a sortie involving winching down of a team of para commandos into an area near Alampil, close to the Brigade HQ at Mullaitivu. News was received at our camp at Vavuniya that this had been deliberated upon and rejected as it was too risky, involving hovering for an extensive period in an area which was known to be hostile and very active with LTTE elements. What later transpired at the JOIR was not known but with a smaller team of para commandos which had to be dropped in a clearing in the thick jungles close to Alampil, where the army was in the process of planning a major operation. The Knights detachment at Vavuniya consisted of four helicopters at that time. The pilots on detachment consisted of eight Flying Officers, of which I was the senior most, with four years of service. [When I was narrating this to senior officer, much later, he asked me which air force was I flying for!] So I made the flying programme accordingly and planned myself with Fg Offr MR Anand for the sortie. We were to be briefed at the Brigade HQ at Mullaitivu on the morning of 23 Jun and undertake the sortie from there. We would be escorted by a Mi-35 from 125 (H) sqn which was based at Trincomalee.

The next morning we took off as planned in Mi-8 (Z-2454) and landed at Mullaitivu airfield. All the aircrew and the para-commando team, with members from all the para battalions, were briefed and we started up for the mission. Someone had a premonition of the events which followed, and group photographs were taken before we got into the aircraft. We reached the small clearing amidst tall trees in the thick jungle about 15 min later, and the Mi-35 carried out two passes to check for any signs of movement of LTTE personnel or any signs of habitation. None were detected and we established a hover over the clearing and commenced winching. Just a few minutes after we commenced this operation, I noticed puffs of smoke emerge from the leaves, heard the sound of small arms fire and a banging sound on the aircraft. I realised that we were being fired at and instructed the gunner to discontinue winching and reel in the cable. This took a few minutes, during which I realised that we were under heavy fire. I could also sense that the cockpit was being targeted at as the acrid smell of gun powder filled the cockpit. All the crew members remained calm and continued to carry out the duties assigned to them. The flight engineer checked and called out that all parameters were normal, the co-pilot that clearance from the trees on the right was sufficient and the gunner continued his commentary on the progress of winching. I gave a call to the Mi-35 orbiting overhead that we were being fired at and commenced take off.

After take-off, I took stock of the situation and asked for a report on the damage. The gunner reported that the leader of the team was hit in the chest and seemed to be seriously injured, and one more member of the Para team was hit in his leg. He also reported a number of bullet holes in the cabin compartment, but no signs of any fuel leaks from the external tanks. The co-pilot and flight engineer received minor injuries in the leg and hand respectively. All other aircraft systems appeared to be functioning normally. We headed straight for the airstrip at Mullaitivu, and on switching off transferred the casualties to the Mi-35, which had landed behind us after emptying all his ammunition at the site of the firing. Unfortunately, it was too late for the team leader, and we came to know that before the Mi-35 could land at the nearest field hospital at Vavuniya, he had breathed his last.

A more leisurely assessment of the damaged helicopter revealed that about 25 rounds had entered the aircraft, three of them into the cockpit. Besides the airframe damage, a number of electrical cables were damaged, and some of the control rods were bent on taking the force of rounds. I still think we owe a lot to Russian Metallurgy, because of which we were still able to fly the aircraft back and land safely. I am also happy to be able to narrate with pride the exemplary and professional behaviour of the other members of the crew, one Flying Officer, one Sergeant and one Warrant Officer, who displayed good team work and continued with the mission without any signs of panic. A BRD team was later sent to carry out a detailed assessment and repair on site, to ferry the aircraft back to base.

A Memorable Experience It was back to business the next day and I continued operating as part of IPKF almost till the end of 1989. We changed tactics at times based on intelligence reports of the LTTE acquiring SAMs. For some time, we were asked to 'hit the deck' and fly low level which we enjoyed immensely, though navigation was a problem over the thick jungles (GPS was unheard of!). For Some time we had to fly above one Km, which was quite a problem as the Mi-8 took its own time to get to that altitude. We flew all types of missions, but we always put in a special effort to ensure that causality evacuation sorties were undertaken without any delay, even at night with makeshift lighting arrangements at helipads. The memory of some of these sorties will always remain imprinted in my mind; sorties undertaken to pick up 10 to 15 casualties at a time, of limbs shattered to shreds in mine blasts, of the dazed look of shock on their faces, of the helicopter dripping with and leaving a trail of blood on the runway as we came into land. We took risks at items, and some of them in retrospect seemed to be foolishly brazen. But we considered them worth taking at that time, to achieve what we had to in those situations. And throughout the operation, the IAF did not suffer any casualties, in spite of the very large number of sorties flown over a long period of time, many of them under direct ground fire, and with so many aircraft involved. I have not come across any records of the numbers involved, but I have heard of an estimate of 70,000 sorties flown over 32 months by helicopters and transport aircraft. I think it speaks volumes about the professionalism of all personnel involved in the operation, right from the beginning to the end.

IN SERVICE OF PEACE: GALLANTRY CITATIONS

Wing Commander Vijay Kumar Narayan Sapre (11285), Flying (Pilot)

Wing Commander Vijay Kumar Sapre was involved in planning and execution of Helicopter Operations in support of Indian Peace Keeping Force (IPKF) in Sri Lanka from the very beginning. On the night of 11th/12th October, 1987, Wing Commander Sapre led the night special heliborne operations comprising of four Mi-8 aircraft. He successfully carried out the entire mission landing the troops at the designated helipad which was under heavy fire by the militants. However, since this was a critical mission, he displayed outstanding grit and courage and completed three successful missions. Though heavy damage was caused to his helicopter, he quickly assessed the situation and flew back the aircraft to base with great skill.

Squadron Leader Thirumangalath Kattil Vinay Raj (11438) Flying (Pilot)

Squadron Leader Thirumangalath Kattil Vinay Raj carried out extensive flying in Jaffna Peninsula during the month of October 1987 as part of the Indian Peace Keeping Force (IPKF) Operation. On the night of 11th/12th October,1987, during the course of night Special Heliborne operation, Squadron Leader Vinay Raj was the Captain of No.2 aircraft in a four aircraft formation, detailed for a special mission. He was responsible for successfully landing three missions at the designated helipad which was under heavy fire by the militants. During this operation the helicopter received extensive damage but Squadron Leader Vinay Raj was able to fly the helicopter back to base safely.

Flight Lieutenant Vishwanth Prakash (17149), Flying (Pilot)

Flight Lieutenant Vishwanath Prakash formed part of the Indian Air Force element of the Indian Peace Keeping Force (IPKF) in Sri Lanka. On the night of 11th/12th October 1987, he was the Captain of helicopter in a four aircraft formation and involved in the critical Special Heliborne operation. Despite being comparatively junior in service and relatively inexperienced, he showed determination and grit of a very high order. He successfully carried out three mission during the night. The helipad was under heavy fire by the militants and his helicopter suffered extensive damage, but he managed to fly it back to base safely.

Squadron Leader Rajbir Singh (14287), Flying (Pilot)

Squadron Leader Rajbir Singh served with the Indian Air Force element of the Indian Peace Keeping Force (IPKF) in Sri Lanka. On the 3rd November,1987 Squadron Leader Singh was detailed to strike the militants strongholds which were impending the advance of a para commando Regiment towards Mulai. On reaching the area he was directed to attack a stronghold which was heavily defended with heavy machine guns. In spite of heavy ground fire, he carried out repeated front gun attacks with deadly accuracy. During his third attack on a machine gun nest the aircraft was hit by ground fire. After pulling out of the attack he noticed that the left engine was damaged and had heavy oil leak. He immediately switched off the left engine. Though the aircraft was heavily loaded with ammunition he decided not to jettison the much needed armament stores. Squadron Leader Rajbir Singh realized that he was flying over a very hostile area and a forced landing would result in captivity by militants. He kept absolutely calm and nursed the aircraft back in a very professional manner. He tried to contact base on radio telephony but the radio set had also been damaged due to ground fire. In spite of heavy traffic over base he landed the aircraft on a single engine without causing any damage to the aircraft.

Flight Lieutenant Nicodemus Manohar Samuel (17449), Flying (Pilot)

On the 22nd Jun 1988, Flight Lieutenant Nicodemus Manohar Samuel was detailed for a special task involving winching down of IPKF reconnaissance party in the most difficult and thickly vegetated terrain in Mullaitivu area in Sri Lanka. This operation involved hovering over on area which was expected to be infested with hardcore militants. During the execution of this mission, his helicopter came under heavy ground fire from the surrounding area. However, with utmost dedication and total disregard to personal safety, he continued to hover till the recce party was retrieved by winching. In the resultant ground fire, the helicopter sustained extensive damage and some of the crew members and the troops on board were injured. In spite of being subjected to heavy ground fire, he kept his cool and continued to fly his helicopter in the damaged condition till his mission was completed. He also displayed excellent airmanship by correctly assessing the damage caused to the helicopter and flying this crippled machine safely to the nearest helipad, thereby avoiding further damage to his helicopter and also render immediate medical aid to the injured troops/crew.

Wing Commander Chandra Datt Upadhyay (11336), Flying (Pilot)

Wing Commander Chandra Datt Upadhyay was deployed as the Type Force Commander of the Pratap Fleet of Indian Air Force Element of the Indian Peace Keeping Force in Sri Lanka from the 19th October 1987 to the 3rd November 1987, when the IPKF was engaged in fierce battle with the militants for control of Jaffna town. On the 19th October 1987 and again on the 20th October 1987, he carried out sorties to take urgent rations, petrol, oil and ammunition for the advancing Army column and evacuate casualties. He carried out the mission by landing on an open field which was under fire from all sides. On the 21st October 1987, a single helicopter special mission of para commando force was dropped at Achuvail and para commandos and prisoners were picked up from Puttur West. During take-off from the ground at Puttur West, his helicopter was fired at by the militants by machine gun & rockets. One rocket damaged the trailing edge of a rotor blade, causing severe vibration. The helicopter went out of control and started descending rapidly. Wing Commander Upadhyay showing great professional skill controlled the helicopter only a few feet above the ground and turned back and landed the stricken helicopter in the same field. All the passengers and crew took shelter in ditches and once the cross fire ceased, he inspected the helicopter, boarded the passengers and managed to fly it back at low speed to the airfield, thus, saving valuable lives and a helicopter.

Flight Lieutenant Atanu Guru (16794),Flying (Pilot)

Flight Lieutenant Atanu Guru was inducted into the Air Force Element of the Indian Peace Keeping Force in Sri Lanka with the very first fleet of Akbars while he was still undergoing conversion. On the 26th October 1987, he flew in a mission ex-Jaffna to search and destroy any militant vehicle operation on the Jaffna Karaitivu causeway. Five vans were located on the causeway attempting to flee towards Karaitivu. With extraordinary accuracy and commendable calmness, he destroyed these vehicles and killed all the militants aboard these vehicles. On the 3rd November 1987, he flew in a close air support mission for the one the Commando Battalions of the Para Regiment. The Commandos were surrounded by the militants and subjected to heavy automatic fire. He engaged the militants locations on the para commando perimeter. These locations were within three hundred meters of friendly forces. The militants engaged the aircraft with automatic weapon using tracer ammunition and scored a hit. In spite of the visible stream of ground fire, he engaged the militants. His action forced the militants to retreat from their positions. The Para Commandos won this action solely because of the air support provided by his determined attack on militant positions.

On the 7thDecember 1987, he flew in a search and destroy mission in support of one of the Commando Battalions of the Para Regiment in area Mulaitivu. The militants disembarked from a tractor and opened fire at the aircraft. In complete disregard of personal safety, he swiftly and accurately returned the fire and silenced the militant's guns. Subsequent firing had destroyed the tractor and killed and wounded several militants. On the 15th December, 1987 while engaging a militant vehicle carrying reinforcements against a battalion of Garhwal Rifles in Mulaitivu area, the militants opened fire and the aircraft was hit on the main rotor blade. Despite this and intense militant fire, Flt Lt Atanu Guru returned the fire and destroyed the van.

Wing Commander Krishan Kumar Yadav (12032), Flying (Pilot)

On the night of 9th November 1987, at 2330 hrs, Wing Commander Krishan Kumar Yadav, task force commander, heliborne operations in Jaffna, was called upon by one of the Division Commander to undertake an emergent mission in which 180 commandos of one of the Para Battalions were to be dropped at Sabaipai to neutralize a strong hold of the militants. Knowing the guerrilla warfare tactics and the prevailing ground situation, he made a prudent assessment of the situation and flew initially a single aircraft with limited troops to the dropping zone to guard against the possibility of large un-escorted heliborne force getting destroyed by ground fire before it would reach the target. His appreciation proved correct. Just after landing, militants took a pot shot on his aircraft. A brave and dedicated leader, he took control of his aircraft and returned the helicopter safely to the base. Undeterred by operational hazards of guerrilla warfare and the fierce ground fire encountered in the previous sortie, he led the remaining two sorties shortly after the first mission and accomplished the task of dropping all 180 troops at the designated place even though the ground fire still persisted.

Wing Commander Dhirendra Nath Sahae (11590), Flying (Pilot)

On the 24th August 1988 when 'OP CHECKMATE' phase (III) was being conducted by 4 Inf Div in Nitikaikulam jungles in Sri Lanka, Wing Commander Sahae was detailed to standby at Mulaitivu helipad with 4 × 16.57mm rockets in a offensive role. During a dare-devil attack on the TAC HQ of 7 Brigade, militants blew up one AOP Chetak helicopter and managed to damage another. They also managed to attack the HQ with RPGs and automatic weapons thus gravely endangering lives of GOC, IPKF Land

Forces, other senior Army and Air Force Officers present at that location. He was immediately asked to scramble in order to deal with the militants threat. He got airborne and reached the area within record time of 12 minutes. He engaged the militants and fired rockets at the area as directed by another airborne Chetak aircraft. He successfully neutralized the threat and ensured safety and security of the TAC HQ and senior Army Force Commander at the site. Three militants were reported killed due to his timely action and they were forced to withdraw without causing any further damage.

Wing Commander Suresh Chander Malhan (11579) Flying (Pilot)

Wing Commander Suresh Chander Malhan (11579) Flying (Pilot) is the Commanding Officer of No. 125 Helicopter Squadron, Air Force since 14th Apr 1986. His Squadron currently forms part of the Indian Peace Keeping in Sri Lanka. The Squadron has been tasked to destroy Liberation Tigers of Tamil Elam (LTTE) strongholds, surface transport and ocean going vessels. He drew first blood in the attack over a causeway south east of Jaffna against LTTE vehicles and an ammunition dump. He spotted the ammunition dump and dived into attack. With his accurate rocket attacks, the ammunition dump was destroyed in a single pass. That was the beginning of sorties of offensive air support missions by his unit. On 27th Oct 1987 he led a major attack on the LTTE strongholds at Chavakcheri. He displayed remarkable marksmanship while attacking in a built up area, destroying LTTE strongholds that had been holding back the advance of the Army. This attack enabled the Indian Army to take over the township with negligible casualties and link up with 11 Madras Regiment which has been isolated for seven days. On 03 Nov 87, 1 Para Commando Battalion was pinned down by deadly enemy fire in the area of Mulai. This was at a time when there were strict restrictions imposed on the use of rockets and bombs. Not a person to be deterred, he struck the enemy positions with front guns. Displaying complete disregard to himself, he made repeated attacks on enemy positions against heavy hostile fire even though his aircraft sustained bullet hits and stopped only when the heavy automatic weapons were silenced.

Flying Officer MR Anand (18257) Flying (Pilot)

Fg Offr MR Anand (18257) F(P) of a helicopter Unit has flown over three hundred and fifty five sorties in Op Pawan till date. Most of these missions involved operating over inhospitable terrain and under most trying operational conditions. On 27th Jun 88, he was detailed as co-pilot for an important but highly dangerous task in the jungle areas south of Alampil, where active ground operations were in progress. However, during the execution of this vital task, the helicopter was hit by ground fire, in which he also sustained minor injuries. Unmindful of it all, this officer in the best traditions of the Air Force continued with the mission and ably assisted his captain in dealing with the situation and recovering the helicopter at the closest helipad without further damage to his helicopter or injury to the troops. Despite his relative inexperience and injury, he displayed courage, determination and professionalism of the highest order.

Flight Lieutenant Ajay Dogra (17145) Flying (Pilot)

On 14th Nov 87, Flt Lt Ajay Dogra was assigned the task of dropping troops and ammunition at a helipad dominated by the Liberation of Tigers of Tamil Eelam (LTTE) guerrillas in Jaffna. The situation was so tense that he had to take off for the mission being fully aware of the heavy machine gun fire around the helipad. He landed there safely and dropped the troops & ammunition and also lifted the casualties; but as the aircraft took off, it came under continuous heavy machine gun fire by the LTTE elements. It was just the skill of pilot that made him manoeuvre the helicopter to safety. By then the helicopter was hit by a volley of shots. The tail boom and the tail rotor shaft were damaged extensively. In spite of being subjected to heavy ground fire, he kept his cool and continued till his mission was completed. He also displayed excellent airmanship by correctly assessing the damage caused to the helicopter and flying this crippled machine safely to the base thereby avoiding further damage and ensured safety of crew and casualties on board.

CHAPTER VII

Securing J&K: Siachen, Kargil and Proxy War

Introduction Any Indian helicopter pilot would readily agree that if one has not flown in J&K in the myriad of roles and operations across this 'heaven on earth', you have not yet tested yourself as a 'chopper' pilot. Pakistan's designs on J&K have ensured that most IAF helicopter pilots have had this privilege to test themselves! This chapter covers the Siachen saga, which continues to date, as also the culmination of Pakistan's strategy of 'bleeding with a thousand cuts' – the Kargil War of 1999. In the gallant riposte that the Indian Army gave, IAF helicopters played a key enabling part, and sometimes a match winning role. In covering only a few key helicopter units, it is hoped that many others who played an equally important role do not take it amiss. The reason to do so is that the chapter allows only this much space. These selected units exemplify the degree of difficulty-nearing impossibility of the tasks, and also bring out all the other aspects of nation-building and humanitarian support that is part and parcel of countering the 'thousand cuts' strategy.

IAF is involved in helping out Border Roads Organisation build roads in the farthest reaches, playing a key role in assisting with railway bridges with the heavy lift Mi-26, helping in repair of electric power lines during snowed out conditions and the list goes on. The role of IAF helicopters in disaster relief during the 2006 earthquake and 2010 Leh landslides are just indicative of the virtually perennial assistance required by the people of J&K. The support to activities such as the annual Amarnath Yatra, support to Election Commission of India, and even to the various task-forces set up by Govt of India to mediate in the peace process, are important milestones in the history of J&K's consolidation into the country. The coverage of the chapter is by no means comprehensive that may require a few volumes; however, it does bring out the focus of the IAF to the joint task in J&K and its commitment to the nation.

The Siachen Saga

That peace is maintained at heavy costs was proved once again when Pakistani intrusions in the desolate high altitude areas of Siachen started in 1984. Such was the nature of the terrain that helicopters were the only lifeline to our gallant troops. The vast glaciated regions are perpetually snow bound and covered with ice. Strong winds, turbulence and extreme sub-zero temperature (-40 degree Celsius) with white-out conditions hamper even routine operations. Helipads are at altitudes of 4.5 to 5.2 km above mean sea level which makes helicopter operations difficult and dangerous with very little reserves of power. More often than not, pilots fly hugging the few valleys and barely clear the ridges which are usually covered with clouds. That the fighting at the world's highest battle field has been sustained over the decades is a tribute to helicopter crews.

Operation Meghdoot was commenced in support of the Indian Army and paramilitary forces in Northern Ladakh to secure control of the heights dominating the Siachen Glacier; also referred to as the world's third pole and potentially a dangerous flash point on the disputed northern borders. IAF Il-76s, An-12s and An-32s transported stores and troops, and airdropped supplies to high altitude airfields; while Mi-17s, Mi-8s, Chetaks and Cheetahs ferried men and material to dizzy heights, far above the limits set by helicopter manufacturers. In fighting for this "roof-of-the world" since April 1984, the IAF's incredible performance at the extremes of temperature and altitude is a benchmark for the world. The Siachen Glacier lies in one of the most inhospitable regions of the world. It contains some of the highest peaks and is one of the most glaciated regions of the world. It is the second longest glacier outside the polar region, its length being 76 km. The area is mountainous, rugged, precipitous and glacial with heights varying from 12,000 to 24,000 feet. It remains snowbound with sub-zero temperatures practically throughout the year. Poor visibility creates problems of navigation, coupled with the high velocity winds, especially after mid-day, and makes flying an extremely hazardous and difficult task.

Ever since it was first demarcated in the Karachi Agreement of 1949, the Cease Fire Line – now called the Line of Control (LoC) – between India and Pakistan, abruptly stops at map coordinates NJ 9842. The LoC was never drawn beyond this point primarily because there has been no military engagement there. Even under the 1972 Shimla Agreement, the LoC was delineated till NJ 9842. Siachen boasts of some of the most challenging peaks in the world, including Sia Kangri, Teram Kangri, Saltoro Kangri, Rimo group and Momostong Kangri. Since foreign expeditions found Indian authorities hesitant to give permission to enter the area, they approached the Pakistanis who readily gave the go-ahead and sometimes logistic support. Hence, some international maps began to show Siachen as a part of Pakistan. Pakistan then unilaterally extended the LC towards the northeast – all the way from NJ 9842 to the Karakoram Pass – completely slicing off 10,000 sq km of Indian Territory, including the Siachen Glacier. Since the Siachen area was accessible from Pakistan occupied Kashmir (POK) through Bilafond La, physical presence of troops in the area during summer was considered operationally essential. On 13th April 1984, Operation Meghdoot was launched. Indian Army permanently stationed over 2,000 men all along the 110 km-long Actual Ground Position Line. They are air maintained by the helicopter and transport fleet of the Indian Air Force.

A Personal Account by AVM MM Bahadur Siachen operations actually started in 1978. Sqn Ldr KDS Sambyal was the detachment commander. We were called to the Div HQ and briefed about a High Altitude Warfare School (HAWS) expedition led by Colonel Narendra Kumar to a glacier called Siachen. A look on the map showed an area where we never thought one would ever be going. The Col GS gave us a short brief on the genesis of the expedition – it was to show the flag on a territory, rightfully ours, but opened to foreign expeditions by Pakistan. Pakistan had also started showing the area as their own on their maps. To oppose this 'Cartographic Aggression', it had been decided to launch the HAWS expedition. The HAWS team that had gone to the Glacier was to be supplied with mail and fresh rations by the Indian Air Force.

It was 20th September 1978 when the first IAF sortie to Siachen was launched. The black snout of the glacier was, to say the least, imposing and menacing yet truly majestic. As we flew along, I looked left and right for force-landing fields. Where there were no crevasses there was only ice. But our Chetak had no skis. Never mind, the army jawan had to be supported and we pressed on. At Camp I, the aircraft was brought to a hover; the sliding door opened and out went the 'fresh'. A little ahead was Camp II and the same thing was repeated. Back at Leh, when we were reliving the experience a thought occurred to us, "Why not pick up their mail for their people back home?" And so, when the request for the next sortie came in on 23rd September 1978, we were prepared.

One could never have imagined that by 1984, when Op Meghdoot was launched, it would become a routine day in and day out procedure, braving enemy fire and that it would be continuing even now! The intervening twenty eight years have seen operations increase in size and intensity. The enemy has tried on very many occasions to dislodge our Jawans from the vantage positions occupied by them – but has had to beat a hasty retreat

each and every time. The Jawans stay in conditions that defy description –
miserably cold, with temperatures going down to –60ºC, where the weather
clamps down for days at a stretch, preventing supplies from coming in. In
such conditions the life-line to the outside world is the Indian Air Force.
Siachen air operations of today are a far cry from the "drawing water from
the well" procedure of 1978. They are scientifically planned and executed
meticulously – for both supplies and human lives are precious. Air
maintenance starts from Chandigarh from where IL-76s and An-32s ferry
in supplies and men to Leh and Thoise. Thereafter, medium lift Mi-17
helicopters air drop loads on to lower level helipads on the glacier – by
lower levels I mean helipads up to about 17,500 ft! Air dropping of supplies
is also done by An-32s at special dropping zones. The Cheetahs then take
over the challenging task of ferrying supplies and men to helipads situated
up to about 20,000 feet.

All operational sorties have an element of risk involved in them,
especially if they are over an inhospitable terrain like the Siachen Glacier.
Pilots have to brave temperatures as low as –60ºC, strong winds, lack of
oxygen and flying in poor weather conditions in close proximity of hills.
There is always the threat of powerful downdrafts that pull the helicopters
down, if the pilots are not careful. Added to that is the fact that the
helicopters are flying at the edge of their flight envelope where the power
margin available is small, if not negligible. Any miscalculation or
mishandling of controls can result in a catastrophic accident. Over and
above, there is the omnipresent threat of enemy fire. The pilots fly fast and
low to give the minimum reaction time to the enemy. The landing on the
match box sized helipads is precise; the Army Jawans open the door, take
out the load, put in the mail/casualty and the pilot executes a take-off – in
the reciprocal direction. The total time on the helipad does not exceed 2
min to 30 sec. The return to the Base Camp is fast, a quick turnaround of
the aircraft is done while the pilots have a cup of tea and the aircraft takes
off for the next mission. Siachen is where only one type of pilot operates –
the brave and a notch above the average; where only one type of technician
succeeds – once again the brave and a step above the average. But what
keeps these pilots and technicians going in such harsh conditions? It is plain
and true "JOSH" and the spirit to ensure that the enemy does not cast an
evil eye on our Motherland.

The 'Twins' in Siachen'   109 HU flying Mi-8s, faced the ultimate challenge
in one of the most unforgiving terrains in the world – The Siachen Glacier.
In Jun 1981, the unit took up supply drops at DZs located at 5 km AMSL
and above. The unit surpassed itself when, in Jun 1984, the CO landed at

5.2 km, firmly establishing the Mi-8 as the world's most versatile helicopter. From Sep 1984 onwards, unit pilots were operating regularly over the Glacier. Landings were routinely carried out at postage stamp sized helipads ranging from 4.5 km to 5.2 km. Thrust into the role of a pioneer, 109 HU became a familiar sight over this vast waste

land. After a year and a half in the Siachen area, during which time the word "Impossible" had ceased to exist, the task was handed over to the Mi-17s closing yet another glorious chapter.

114 Helicopter Unit

Siachen Pioneers in Op Meghdoot (OPMD)   A saga of true grit, determination and sheer persistence has been written by 114 HU. The Unit's legacy has been built over a span of 48 years with sweat, blood and glory. It was formed at Leh which is one of the highest airfields in the world on 1st Apr 1964 with ten Chetak helicopters. Later in the year, it was shifted to Jammu from where a detachment each operated at Srinagar and Leh. Subsequently after the 1965 operations, it shifted to Srinagar where it operated till May 1975, before shifting back to Jammu. In Aug 1987, it moved back to Leh and has been operating Cheetah helicopters since then which are now being systematically replaced by the Cheetal helicopters.

During the second decade of its existence, it had to undergo a change of location from Srinagar to Jammu in May 1975, while detachments at Srinagar and Leh were operated regularly. It was during this period that it shouldered the responsibility of development of this remote region by providing vital transport support and communication facilities. In the extreme winter season and in the face of natural calamities, it was the lifeline of this region. The occasions of providing aid to civil authorities and daring casevac missions from altitudes of 13000 to 19000 ft AMSL are in hundreds. Some examples, in 1977 it undertook the evacuation of a Swiss lady from Nunkun Base Camp (18500 ft AMSL). Massive flood relief operations were carried out in Ladakh region in the same year. In 1976, it again undertook extensive flood relief operations in Ladakh Sector. In 1979, the unit was engaged in extensive mercy missions in the avalanche affected areas in Himachal and J&K sector. Spanish and Japanese mountaineering teams were evacuated from Nunkun Base Camp in July and Aug 1981 respectively. In Aug 1981, a Japanese climber who was critically ill was evacuated from the range itself. In Jan 1982, a Swiss couple, stranded in the Zaskar valley with severe frost bite were evacuated. The unit provided air support to the first ever Indian Army expedition to the Siachen Glacier lead by Lt Col N Kumar thereby earning the name *"SIACHEN PIONEERS"*.

1984-2000   It was evident by 1984 that our territorial differences with Pakistan over the extreme northern frontiers could only be resolved militarily. The launching of Op Meghdoot in April 1984 heralded the involvement of 114 HU into the realms of air maintenance, casevac and communication over the most difficult and inhospitable terrain. The first landing at Siachen glacier was carried out by a Chetak helicopter of this unit on 24th Apr 1982. It was soon realised that Chetak helicopters were not at all suitable for such high altitudes and the unit was fully re-equipped with indigenously built Cheetah helicopter in Aug 1987. It was relocated at its current location on 11th Aug 87 so as to facilitate better inter-services cooperation between the army and the IAF. The primary task of the unit was performed initially by maintaining a Dett each of three helicopters at Thoise and Base Camp. These helicopters use Base Camp as launching pads for carrying out air maintenance in the Northern Glacier, Central Glacier and Southern Glacier (Earlier known as Main, Side and southern glacier respectively). Since then, on an average, annually, the unit has been achieving a load drop of over 500 tonnes in the glacier.

To make the matter even worse, the heavy enemy fire and close proximity of helipads to the enemy posts rendered each and every mission very dangerous and highly demanding. Enemy firing continued throughout the 1990s; and, on 26th Oct 1997, the first of the incidents took place where a helicopter was damaged due to enemy firing. In another incident, on 18th May 1998 the aircraft was put down at Amar helipad by pilots due to a technical snag. An aircraft was launched from the Base Camp for the rescue of the unit pilots. Though the aircraft at Amar was completely damaged due to enemy firing, the rescue pilots were able to land at the helipad in the middle of heavy shelling from the enemy and get their comrades back safely, once again showcasing the valour which is expected from every 'Siachen Pioneer.' A couple of months later, on 6th Sep 1998 while offloading rations, an enemy shell exploded 15 ft away from the aircraft causing heavy damage to the port door and radio panel. Another incident where an aircraft was damaged during enemy fire occurred on 8th Jun 1999 when it was hit by splinters from an exploding artillery shell. 114 HU became the first Chetak/

Cheetah unit to receive the prestigious President's Standard on 13th Nov 1996.

Siachen Pioneers greeted the new century with a declaration of ceasefire from both the countries after Op Vijay, but that did not stop the air maintenance in OPMD. The unit was the first helicopter unit in IAF to receive a Citation by the Chief of Air staff on 8th Oct 2004. It set a new world record on 2nd Nov 2004 by carrying out the highest landing at Saser Kangdi peak at a density altitude of 25125 feet. The tasks carried out by the unit were not only appreciated in the country but interested international ears too; the unit's efforts were recognised for the *Longest, Highest and Most Daring air maintenance operations in the world in 2007 by Guinness Book of World Records.* With ageing of the Cheetah fleet, a need of a better aircraft in terms of efficiency and performance was felt necessary. HAL came up with the new re-engined Cheetah and trials commenced from 2007. This aircraft was as good as the old helicopter except that it had more efficient engines. A beginning of a new era was marked on 12th Aug 2009 with induction of three of such helicopter now known as Cheetal. The effort put in by the unit in sustenance of the army troops deployed in Siachen Glacier can be gauged by the quantum of the flying and load achieved. In 2011-12, it achieved 594 Tons of load and flew 1598 hours for OPMD. Details for the last ten years are summarised below. A comparison of its task and achievements with the numerous army aviation units in Leh brings into perspective the commitment and dedication.

Period	Total Load Allotted	Avg Load Allotted Per Year	Total Load Achieved	Avg Load Achieved Per Year
2002-11	4980 Tons	498 Tons	5054.1 Tons	505.41 Tons

In comparison, Army Aviation (all Flights) details are summarised below:

Period	Total Load Allotted	Avg Load Allotted Per Year	Total Load Achieved	Avg Load Achieved Per Year
2002-11	4300 Tons	430 Tons	3350 Tons	335 Tons

Leh Flash Floods   On the night of 5th-6th Aug 2010, Ladakh experienced unprecedented rainfall leading to flash floods and mud slides. Lines of communication were disrupted and road network severed. This period is also the peak season of foreign tourists. Due to the calamity and its intensity, hundreds of tourists got stranded in remote corners of Ladakh. Seven helicopters mission was launched to evacuate **eighty one** foreign and seven Indian tourists from Skyu village. The mission was critical in terms of restricted space, no landing sites, poor valley clearance and altitude above 12,000 feet. On 11th Aug, once again a similar mission was launched to evacuate twenty three foreigners from 14,000 feet. On 18th, it was tasked to evacuate the bodies of two foreign nationals from Sumdo Chenmo village located at an altitude of 13000 ft. These remains had been submerged in water for twelve days getting highly decomposed. In this mission helicopters were required to be flown above 14000 ft with doors removed and landing in restricted area.

The operating heights of 15000 to 25000 ft (density altitude) are outside the original performance graph supplied by the manufacturer. Not only do these regular operations involve tricky landings at very small makeshift helipads on snowbound table top/saddles, pilots also have to counter the vagaries of inclement weather with temperatures between +30°C to –50°C. The aero medical aspects of extreme cold and hypoxia are part and parcel. Furthermore, white out conditions, scanty visual references, severe clear air turbulence and marginal reserve of power with no go round options demand a very high degree of skill and professionalism by the aircrew. Much has been said and written about these operations but without a mention of our technical tradesmen the saga shall remain incomplete. To service, rectify and maintain the aircraft under such extreme condition is indeed a herculean task, especially so when the average monthly flying task of the unit ranges over 300 hours. With innovation and skill, targets are achieved without compromising on flight safety. The grit, courage and determination showed by all are unmatched. It is not an understatement to state that the unit will be completing three decades of operations on the world's highest battlefield of Siachen Glacier, probably a record unparalleled in the annals of military aviation. While serving the nation with honour, blood and sacrifice, Siachen Pioneers have lived up to the unit motto *'we do the difficult as a routine; the impossible may take a bit longer.'*

Aircraft Maintenance in Sub Zero Temperatures

Martyrs of 114 HU

S No	Name	Date
01	Sqn Ldr AK Dogra	**03 Nov 1978:** Spacing Cable Rupture; 25 km west of Leh.
02	Sqn Ldr Monga	
03	Fg Offr Ghatge	**26 Nov 1990:** Eng Failure due to rear bearing failure at Charasa village
04	Fg Offr K Naresh	
05	Flt Lt Juyal	**02 Jul 1997:** Bad Weather.
06	Flt Lt Agarwal	
07	Flt Lt Kagdi	**04 Jul 2000:** Mixing Unit Failure between Saser Brangsa and Gapshan.
08	Flt Lt Puranik	
09	Flt Lt Krishnamurthy	**03 Oct 2000:** Mixing Unit Failure, south of Kumar helipad.
10	Sqn Ldr SM Bhardwaj	**16 Aug 2002:** Engine Failure Initiated by sand contamination in Governor.
11	Flt Lt M Trikha	
12	Sqn Ldr S Basu	**11 Apr 07:** JPT shoot up and RPM drop,; crashed on Paki side of Amar.
13	Flt Lt Amit Sharma	

Gallantry Awardees

SI No.	Rank	Name	Award	Year
1	Flt Lt	B Ramesh	Vir Chakra	1971
2	Fg Offr	S S Dhillon	Vir Chakra	1971
3	Wg Cdr	JS Kaahlon	Shaurya Chakra	1981
4	Flt Lt	BD Singh	Shaurya Chakra	1982
5	Sqn Ldr	KPN Singh	Shaurya Chakra	1984
6	Sqn Ldr	RS Tandon	Shaurya Chakra	1984
7	Sqn Ldr	SK Dixit	Shaurya Chakra	1984
8	Flt Lt	M Singh	Vir Chakra	1990
9	Sqn Ldr	RR Srinivas	VM Gal	1996
10	Flt Lt	SS Pendse	VM Gal	1997
11	Flt Lt	V Chaturvedi	VM Gal	1997
12	Wg Cdr	VV Bandopadhyay	VM Gal	1998
13	Flt Lt	JVS Kumar	VM Gal	1998
14	Flt Lt	Bhupinder	Vir Chakra	1998
15	Flt Lt	D Ganguly	VM Gal	1998
16	Sqn Ldr	HK Sachdeva	VM Gal	1999
17	Flt Lt	RS Kagdi	Shaurya Chakra	2001
18	Flt Lt	AS Geharwar	VM Gal	2001
19	Flt Lt	RC Pathak	VM Gal	2001
20	Sqn Ldr	SK Bhatnagar	VM Gal	2004
21	Sqn Ldr	Bhale Rao	VM Gal	2004
22	Sqn Ldr	S Basu	Shaurya Chakra	2008
23	Flt Lt	A Sharma	VM Gal	2008

AN EXAMPLE OF TRUE & SINCERE JOINTMANSHIP

Col Kaushal Sreedharan
Director AA-4 (Flight Safety)

Tele : 011-25694697 *(O)*
              36809

00563/FS/DO/AA-4

Additional Directorate General
Army Aviation
General Staff Branch
Integrated HQ of MOD (Army)
DHQ PO New Delhi-110011

07 Jun 2012

Wg Cdr UKS Bhadauria
CO 114 HU
Pin-937114
c/o 56 APO

Dear Air Marshal,

1.   I am writing this DO to laud the prompt and exemplary action taken by your pilots, Sqn Ldr Sunil Agnihotri and Flt Lt Girish Boldara in evacuating Maj Amit Mohindra and Maj CS Singh from a forward post in the Northern Glacier post their crash on 23 May 2012.

2.   The efforts of these two professional and dynamic pilots have helped Maj Amit Mohindra in effecting a near complete recovery from his serious injuries.

3.   I am sanguine that such camaraderie and jointness in execution of operational tasks will go a long way in fostering true jointmanship in both thought and action.

With warm regards,

yours sincerely ,

Copy to :-

Air Cmde SP Wagle, VM
AOC 21 Wg AF
Pin – 937114
c/o 56 APO

AVM PN Pradhan
ACAS (Ops) T&H
Room No – 554 A
Air HQ, Vayu Bhawan
Rafi Marg,
New Delhi - 110106

PD (n)

19/c.

PD Ops (.H)
Dy No...............
Date................

130 Helicopter Unit

The Condors 130 Helicopter Unit, christened as the Condors, was formed at Jammu on 15th Feb 1988. Six Mi-17 helicopters were ferried in from the Aircraft Erection Unit (AEU) Bombay with the Government of India sanction on 6th April. Equipped with the TV-3 117 MT turbine engines, they proved to be a helicopter pilot's dream, both in terms of power as well as handling. 'Rana' as the Mi-17 was named by the Indian Air Force, lived up to its name by proving itself as a valiant workhorse. The rotors were soon to touch the azure blue skies across the country, from the frozen frontiers of the Siachen Glacier to the verdant rain forests of the North East. In addition, they were also destined to strike terror and ensure peace in the emerald kingdom of Sri Lanka. The Condors drew first blood with the aircraft being inducted into the deserts of Rajasthan, the very next day of their induction, to take part in Exercise Hammer Blow.

A Flying Start A nascent unit in a new air force station flying pristine aircraft would have involved time to settle down into a semblance of administrative and operational routine. When the area of operations is the highest battlefield in the world, there are phenomenal hurdles to surmount; yet, all personnel went about effortlessly tackling these teething hitches. That the unit flew 50 percent of its total flying, 1338 hours just in the OPMD area airlifting 1584.390 tons in the operational area of the Siachen Glacier during the first financial year i.e. from Mar 1988 to Apr 1989, speaks volumes of the kick start it got. It shared the Herculean task of air maintaining the Indian Army in Siachen Glacier in Operation Meghdoot along with helicopter Units from Sarsawa, Hindon and Udhampur during alternate months. Mi-17s were positioned at Thoise (Transit Halt of Indian Soldiers Enroute) which also happens to be the northern most airfield in the country. Away from Siachen, Condors continued to be in full readiness for employment in the armed role against ANEs and continued air support in other parts of Jammu & Kashmir. From the modest start of five helicopters and 61 airmen, it gradually grew in strength to 10 helicopters by Sep 1988.

The hard work and commitment of the Condors started bearing fruit on the 26th Jan 1990. The award of five gallantry awards/commendations was indeed a resounding recognition of its arrival. Statistics stated 1575:45 hrs in OPMD while airlifting over 2003.434 tons of supplies during the period of Apr 1989 to Mar 1990. With their capabilities well documented, it was only a matter of time before army exercises started trooping in. There was a spate of these during 1990 with the unit participating in HESU activation as well as in exercise Vajra Vijay from Adampur. In 1991, the Condors crossed the magical figure of 10,000 hrs by September; and flew a staggering 2409.50 hrs in a single year.

Milestones　　With mounting confidence in the mettle of both the men and the machines, increasingly difficult tasks were being allotted. It started with a Special Heliborne Operation exercise codenamed Sangharsh, where 158 Para Commandos were trained for the role. This was soon followed by the massive relief operations undertaken in the wake of a devastating earthquake in the Uttarkashi area. The unit helicopters were instrumental in saving numerous lives by timely evacuations. From Uttarkashi, the Unit was to fly headlong into Exercise Prasikshan, where underslung operations in Mi-17 aircraft by night were undertaken for the first time. This was followed by a mission to airlift an L-70 artillery gun from Kargil to the Indian Army post named 'Chhoti' on the Line of Actual Control (LAC) in the Kargil sector. The task was a challenge to both man and machine. The 'Chhoti' post at an altitude of 4 km above mean sea level is situated in the line of sight of enemy positions, which were certainly not to take very kindly to any qualitative change in the force level and fire power of the other side. In addition, air lifting an artillery gun to a confined area at an altitude of 4 km takes man and machine well beyond the safe flight envelope. The task as usual, was undertaken readily and executed with professionalism. The benchmark of 15000 hours was crossed in 1993, and it went on further to fly 443 hours in a single month of Jun 1993, thereby establishing standards which were bound to stretch the capabilities of peer units trying to emulate it. They went on to cross the 20,000 hours mark without much ado in 1995.

Supreme Sacrifice　　In the month of Aug 1996, tragedy struck the nest with the painful loss of four brave Condors who laid down their lives in the service of the nation. Recounting the sequence of the events the citation had this to say:

> "In the month of August 1996 Flight Lieutenant Sandeep Jain and Pilot Officer Vaibhav Bhagwat were detailed for crucial air maintenance in OP Meghdoot area. During the period, the enemy had redeployed the forces in the Southern Glacier area and had established new bunkers threatening our posts in the area. Accessibility to our posts had also become extremely difficult since helicopter routes to the posts were under the enemy's constant watch and firing range. In this situation air maintenance of our posts was almost impossible. It was decided to launch a few dedicated air maintenance sorties to maintain the crucial post at Hoshiar. Flight Lieutenant Sandeep Jain was detailed as the Captain of these missions. Pilot officer Vaibhav Bhagwat, Sergeant Rudresh and

Sergeant Jha were the rest of the team. They were thoroughly briefed about the risk involved. Notwithstanding the risks, the dedicated crew under the command of Flight Lieutenant Jain undertook the challenging task bravely and intelligently. Two such missions were carried out successfully in spite of enemy's interference. During the third sortie on 26th August 1996 after the load was dropped, their helicopter was engaged by the enemy ground forces and shot down. Flight Lieutenant S Jain and his crew died in this accident."

Operation Rakshak-III (J&K CI Ops)   Counter-insurgency operations by the army, paramilitary, air force and police forces in Jammu and Kashmir State are termed Operation Rakshak III. Operations Rakshak I and II were conducted by the army in Punjab during 1990-91. Initially the army's role in J&K was limited to curbing infiltration from the border, as in the Punjab, while the insurgents were fought by the J&K Police, the CRPF, and the BSF. However, when the extent of Pakistan's involvement became clear and a proxy war was recognised, the Indian Army stepped in to lead the fight. 130 HU has been instrumental in providing operational support to the ground forces by providing vital tactical airlift capabilities. It has been a pioneer in providing logistical support to the ground troops stationed in the sector of J & K especially during the time of severe winters. Special operations with helicopters allow lethality and advantage of surprise in operations. This Unit has been regularly carrying out operation in co-ordination with the ground forces in order to curb terrorists operating in the valley and preventing them from causing any major catastrophic incident. Condors have been carrying out regular operation in almost all the sectors of the J & K namely Delta, Romeo and Zulu. In one such instance, 21 terrorist were killed after 130 HU inducted troops in Baraub, which could have resulted in major terrorist activity in the valley. In another instance, in 2010 when some suspected movement was picked up by a UAV on the ridges behind Amaranth temple, 130 HU was tasked to induct troops in close vicinity, which was executed successfully.

A Sample of Tasks in OPMD

- Apr 88 – Mar 89 1338:35 hrs 1584.390 tons
- Apr 89 – Mar 90 1575:15 hrs 2006.434 tons
- Apr 90 – Mar 91 1199:15 hrs 1668.093 tons
- Apr 91 – Mar 92 1621:56 hrs 1848.246 tons
- Apr 92 – Mar 93 1983:25 hrs 1864:940 tons
- Apr 93 – Mar 94 1890:25 hrs 1977:135 tons
- Apr 94 – Mar 95 1140:40 hrs 1301:025 tons

- Apr 95 – Mar 96 1146:20 hrs 1189:350 tons
- Apr 96 – Mar 97 1192:15 hrs 1398:146 tons
- Apr 97 – Mar 98 1031:05 hrs 1864:940 tons

Op Safed Sagar (May-Jul 1999)   The early months of 1999 saw the Pakistan Army's attempt at gaining hold in Jammu and Kashmir under the garb of 'mujahedeen', an operation very reminiscent of their venture of 1947. However, this time locale chosen was the snow covered heights of the Kargil, Drass and Batalik region. Besides trying to internationalise the Kashmir issue the endeavour was to dominate the strategic national highway 1A and isolate Leh/ Siachen glacier. The Indian army was immediately tasked to evict the unwanted intruders. The infiltrators had established themselves in positions of eminence on high ridges. This was difficult to neutralise by ground action alone. The IAF was first approached to provide air support on 11th May 1999 in the form of helicopters. "Go ahead" was given on 25th by the CCS to the IAF to mount attacks on the infiltrators without crossing the LC. Thus the IAF joined the fray on 26th May 1999. These operations included rocket attack, communications, air logistics and casualty evacuation. IAF Mi-17s were called upon to take on the well entrenched enemy with its lethal firepower of 64 mm rockets. The men and machines of 130 HU also provided the helicopter effort during this operations flying a total of 215 hrs and lifting 715 tonnes of load a substantial effort, considering the altitude of the area. This effort was commended by the President of India when he bequeathed the Unit with 'Battle Honours' for the Unit's admirable performance during the operations.

Op Sarp Vinash (2000)   The Indian Army's Northern Command conducted a complex militant camp-busting operation called Operation Sarp Vinash with skill and precision; easily one of the landmark counter-terrorism operations in Hill Kaka area of Jammu & Kashmir. Hill Kaka was no Kargil in its strategic importance but merely a staging post for Pakistani militants. 130 HU was called in to provide air support tactically and in terms of logistics. It was estimated that up to 100 militants were in and around the hideout spread out in the forest when Special Forces carried out the initial raid. They killed 13 Pakistani militants and in subsequent combing operations which lasted 10 days, seven militants were ambushed near Haripur while they were attempting to cross over into Srinagar. Altogether, 45 militants were killed against a loss of four soldiers killed and two wounded. Substantial recoveries were made: approximately 60 caches and hideouts were busted yielding 20 AK 47 rifles, 5 PIKA guns, two sniper rifles and unspecified quantities of grenade launchers, self loading rifles and 45 kg of plastic explosives. In addition, substantial quantities of radio sets and other communication equipment were also recovered besides enough rations to feed 500 men for two weeks.

When the first phase of Operation Sarp Vinash began, the first thing that Romeo Force did was to construct three helipads in the region, at heights between 10,000 and 11,000 feet, which were utilized to form the logistics chain in terms of man and other logistical support to carry out operations under various constraints, including the inability to deploy troops on a long-term basis due to lack of access. Op Sarp Vinash has been instrumental in establishing peace and stability in the entire Rajouri Sector; this success would not have been possible without hard intelligence and tactical airlift by the IAF.

In Harmony with the Indian Army & People of J&K   Like all IAF HUs, 130 HU also has to continuously strive to meet multifarious demands that never end. Multi-tasking and multiple skills is the hallmark of IAF helicopter pilots; and, Condors are no exception. The sample of tasking in the paragraphs below will clearly bring this aspect to all.

Oct 2001 – Dec 2003   Under Wg Cdr KM Reddy, it flew extensively in J&K sector including Romeo, Delta, Srinagar, Kargil and Siachen glacier. It also carried out extensive flood relief operations in Feb 2003 rescuing soldiers as well as civilians. A dozer was airlifted in the sensitive JWG sector weighing 2280 kg. Also, a large number of civilians were evacuated due to heavy snowfall in Ladakh from Thoise and Padam in Mar and Apr 2003.

Jun 2004 – Jul 2006   Condors operated extensively in the J & K sector which involved air maintenance of Delta, Romeo, Warwan, Kargil and Siachen Glacier. Airlift was also provided to the J & K civil administration due to a number of natural calamities in this period including heavy snowfall, avalanches and an earthquake. It carried out extensive missions to ensure peace in the state by carrying out various counter insurgency missions in partnership with the Unified Command. The Unit also carried out extensive flood relief operations in Himachal Pradesh, Madhya Pradesh and all the way down to Gujarat.

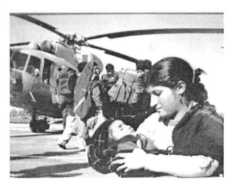

Jul 2006 to Mar 2008   It carried out extensive operations in Delta, Romeo, Srinagar, Kargil sectors other than air maintenance of the Siachen Glacier. It made a record of sorts by dropping 56.5 tons of load in a single day in the

Glacier in Mar 2007. The counter insurgency operations carried out in August in Banihal and Kishtwar sector received a lot of media coverage. In Oct 2007, an operation in Warwan sector to help civilians in a village which had burnt down continued well into the night. Extensive relief operations were carried out in Feb 2008 after the heavy snowfall in J & K. This was in support of the civil populace and involved areas of Kishtwar, Navapachi, Sondar, Kupwara, Tangdhar, Padam, Lingshet and Kargil. 246 passengers were rescued from Leh sector.

Mar 2008 – Mar 2011   It continued its task of air maintenance of the Siachen Glacier and Delta, Romeo, Kargil and Srinagar sectors. 591 passengers were evacuated from Warwan sector in Dec 2008. In Mar 2009, it participated in Ex Parvatarohi; 45 troops were inducted and de-inducted at Chungtash at a height of 4.6 km entailing a herculean effort. In Apr 2009, extensive support was provided for election duty for J&K and airlift of media personnel. 149 passengers were rescued from Kargil-Leh-Kumathang-Kargil utilising 43 sorties. On 30 Oct 2009, during a BAS mission in Doda valley, one helicopter of the Unit crashed in the river killing all four crew on board; the operations continued. A number of rescue operations were undertaken which included rescue of stranded civilians on the Tawi river bed during floods, rescue of 28 civilians near Kathua and the enormous rescue effort during the flash floods in Leh in 2010. Condors had a lead role in counter-insurgency operations missions, especially during the preparations for the Amarnath Yatra. 1235:10 hours were flown in 2010-2011 which included 1049:40 hours towards OPMD.

Apr – Dec 2011   A quantum jump in rescue operations and air maintenance during this year was witnessed; 1200 tons of load in air maintenance in Op Meghdoot sector alone flying over a 1000 hours till Dec 2011. The unit carried out slithering of 116 troops during Ex Gauntlet in the first quarter. Ex Vijayebhav saw slithering of another 132 troops during a mission. Casevac of one army patient was done during this quarter. During Jul –Sep 2011, it carried out air maintenance in the Gurez and Delta sector airlifting 29.15 Tons and 97 army personnel. Casevac of four lying patients of army and 45 of J & K including 30 lying patients was a highlight. Rescue operations were carried out in the Surankot area with six people being rescued from a river bed involving a herculean effort of a continuous sortie of 2 hours 30 min. 192 troops were trained in slithering ops including 128 troops by night. 16 troops were trained in STIE Ops while 64 troops of 6 MAHAR were trained for slithering. During the last quarter of year 2011, rescue operations saw rescue and airlift of 13 army patients and 34 of J & K govt including 16 lying patients. 158 troops were trained in slithering in Riasi area, 128 troops in Op Kamyab in Amritsar and 200 troops from Udhampur. Apart from regular operations in the Siachen, it undertook a daring drop of essential supplies at a post named Chaman in the Glacier; taking the limits of men and machine to all together new heights and keeping the flag of Indian Air Force flying high.

126 HU-The Featherweights

The continuing instability along the borders, especially in the Siachen area gave rise to the requirement of heavy lift logistic support capability. Consequently, it was decided to induct the Mi-26, the giant heavy lift helicopter of Russian origin in May 1986. The Mi-26 is the world's biggest helicopter of conventional design and has a capacity to carry almost the same load as large fixed wing transport aircraft. The unit got down to business in good time after completing training in the erstwhile USSR. The helicopters were brought by sea and erected at Bombay before being flown to their permanent locations; in itself a major achievement. Within months of induction, the unit created a record of sorts by landing this huge machine at DBO – a helipad located at some 5000 meters. In Aug 1987, the unit airlifted 16 hospital containers from Leh to Thoise enabling the Army to make a hospital fully operational at these forbidding altitudes. In 1988, sensitive computer equipment that could not be moved by road was airlifted from Delhi International Airport and delivered at the very doorstep of an institution. Since then it has been rendering yeomen service to the Indian Army in J&K. It has helped civil authorities to achieve the impossible in far-flung areas of Northeast and the entire Himalayas—lifting road building materials, bridges, heavy machinery for dams—and the list goes on. It is a great force-multiplier in war as a heavy lift quick reaction mobilisation platform.

153 HU: Daring Dragons

The icy wastelands of Siachen Glacier have been a bone of contention between India and Pakistan since independence. It was only in 1984, that the Govt of India alarmed by the frequent incursion by Pakistani troops decided to extend their control over the glacier by moving in the Indian Army in occupy those dominating heights. The inhospitable terrain and the geographical location of the Siachen Glacier make it impossible for any suitable mode of transport except a helicopter to operate and maintain our troops in that area. The Daring Dragons ventured into uncharted territory and untested waters on 10th Feb 1989 within four months of being formed as part of the Indian Army's efforts to secure the Glacier under the code name of Operation Meghdoot. The Siachen Glacier poses its own set of problems where both man and machine are put to the ultimate test, Braving sub zero temperatures, Blizzards, Snowfalls, lack of oxygen and high altitude sickness the officers and men of this brave fighting outfit have completed over 15 yrs of yeoman services to the nation in the Glacier alone. The machines are also flown to their limits many a times beyond the service ceiling, as laid down by the manufacturers. Helicopters operate under constant threat of enemy shelling and attack by IR guided MANPADS. In Oct 1990, Sqn Ldr Y Rajora's helicopter was hit by enemy small arms fire damaging the cabin floor and rotor blades; however the helicopter was safely flown back to base. Braving all the odds the unit has continued to work firelessly year after year having flown tens of thousands of hours in OP Meghdoot in air dropping rations, kerosene oil, mail etc.

Op Sarpvinash   Flushing out militants from J&K reached its high point when the unit was tasked to lend a helping hand to the various military outfits engaged in this was of terror in the hills of Pir Panjal. In Jun 2003, a number of militant hideouts were busted and the so called safe havens came crumbling down with the relentless and persistent air effort provided by the Dragons. Since then, the posts are air maintained and the Dragons have been flying in this area on a routine basis. Flying in these areas with armoured protection, loaded CMDS racks and heat shield suppressors has become a matter of routine. While others only fly and navigate, the unit pilots are also in the lookout for any plumes from an IR guided missile or the lurking threat from small arms fire, during critical phases of takeoff and landing in this multi spectral theatre. The unit has devised methods to beat the ANEs at their own game by flying beyond missile envelops, adhering to random routings or just skimming tree top levels to ensure safety. All this requires professionalism, courage, determination and skills of the highest order.

Relief and Rescue   The first two months of the year 2005 tested the resilience of our great nation as never before. Close on the heels of Tsunami, the state of J&K has braved the onslaught of unabated, widespread and unprecedented snowfall. The mercury plummeted and the road link to the valley remained severed. Incidents of avalanches were rampant and caused widespread damage to life and property. Qazigund area and remote villages in Doda, Kishtwar and Sekhlu sector had been especially hard hit. The intensity of IAF helicopter operations demonstrated a dedication and cooperation between IAF, civil polity and bureaucracy like never before. The first phase saw an air bridge being created across Banihal wherein troops, casualties and stranded civilians were airlifted from Srinagar to

Jammu and Udhampur. One of the highlights was the swift evacuation of as many as 600 civilians by IAF helicopters. A large number of them were ladies and children, marooned on account of blockage of the National Highway. This was done by establishing an air bridge across the choke points on the highway from Banihal to the safety of Chanderkot. It was achieved expediently within few hours in spite of inclement weather. Likewise 1200 stranded civilians in Banihal sector were air rescued to Srinagar by IL-76 aircraft in a single day.

128 Helicopter Unit

Siachen Tigers   From the beginning, the men in the unit had been handpicked for the task at hand. The unit boasted of the very best among helicopter pilots and under the leadership of 'Mike' Dutt and Rama, all pilots were ready for the impossible. In the very first year of its formation, it headed north in April 85 to tackle the greater heights across the Khardung-La on the moonscape-like, bleak and desolate terrain of the highest mountain range of the world. The unit stayed at Thoise and flew all day long. Eight to ten sorties a day was the average rate per aircraft, and it was a new experience climbing up and air dropping supplies at altitudes as high as 19000 ft. They took the aircraft to the limits of its flight envelope and sometimes even beyond. The designers from the 'Mil Bureau' would have been extremely proud of their creation had they witnessed the aircraft in action. Like all good aircrafts have, the Mi-17 also forged a strong bond with its pilots. The pilots on their part were extremely proud of having the privilege of flying

this versatile and formidable machine. The living conditions, to put it mildly, were tough but everyone took it in their stride. All that surfaced was some good natured grumbling at the bar. The sustenance of Indian Army in the unforgiving barren lands of the glacier depended on the acts of these men and their machines, and they lived up to the expectations placed on them. Siachen Tigers flew all day and every day, and flew hard and flew well.

Artillery Guns Beyond Bilafond-La   The Indian Army wanted heavy artillery guns to be airlifted to posts beyond Bilafond-La, a pass over 17000 ft high. These posts were being shelled by the enemy incessantly and the weather was marginal. Hanfee and crew got airborne and airlifted the guns part by part to the posts. The shelling continued throughout the day but the guns were delivered, assembled and were made ready for use within a day. This single feat uplifted the morale of our Jawans and there was a befitting reply to every shell which fell on our side. This was a record of sorts since such a task had never been undertaken before. Hanfee was awarded a Vir Chakra while the other two received 'Mention in Despatches'.

Op-Pawan   In 1987, hostilities broke out in Sri Lanka. Siachen Tigers were at the forefront of the operations and soon found themselves in the thick of hostilities. Flying in Sri Lanka involved undertaking operations under hostilities never encountered before by the Indian Air Force. The enemy was well merged with the normal populace and posed far greater danger than regular troops. The limitations imposed by unfamiliarity of the terrain, weather conditions and lack of navigation aids, including proper maps, made the operation highly challenging. Helicopters came under hostile fire on several occasions and continued to operate in an increasingly antagonistic environment. The high standards of training, meticulous ground work and unparalleled experience of the Siachen Tigers in the highest battlefield of the world, held them in good stead. And as the say "If you have flown on 'the Glacier', you can fly anywhere in the world".

IAF Helicopters in Counter-insurgency

Counter-insurgency operations in J&K have been a prime focus of the Indian Armed Forces. This effort is further supported by various central and state forces, for many years now. These operations in J&K have been termed as Op Rakshak-III. Initially the armed forces' role in J&K was limited to curbing infiltration from the border, however when the [external] state-sponsored anti-national elements (ANEs) flourished in the region, it became clear that a proxy war was on. The raising and deployment of various forces in the J&K sector were made possible with tremendous helicopter support. They took the lead for quick insertion and extraction of Special Forces, and armed role in many time-sensitive missions. The Mi-17 class of helicopters have become a prerequisite for any success.

The area extends from south J&K including Delta, Romeo sectors up to northern edge of LOC along Gurez sector. Mi-17s have been undertaking SHBO, STIE and various other operational support missions for more than a decade. These units have also undertaken hundreds of special operations from 2003 to 2008. Mi-17 helicopters have been the backbone of air power employment in this low intensity conflict. The missions mentioned below are just a few from the plethora flown by the IAF in the region. The spirit of service before self and deep rooted sense of devotion and professionalism of the IAF air warrior has been at the root of achieving these missions with safety. These examples of jointmanship towards a larger cause are a role-model for the future.

Op Rakshak (1990-91) Towards Op Rakshak III, IAF Mi-17 units were tasked to induct troops in Baraub sector. Due to the swift action and retention of surprise, a group of 21 ANEs was intercepted and killed in this operation. In order to give a glimpse of the kind of operations undertaken by the Mi-17s in the region, a few are enumerated in the succeeding paras.

29 Nov 1998 Three sorties were launched from Udhampur, Srinagar and Baramulla to drop Special Forces to Point 3640 in Uri sector. The objective was to take over the post called Garaza Gali. The ANE control of this post was helping induction of terrorists in Uri sector. A total of 60 commandos were to be dropped by two heptrs of 153 HU and 129 HU. The heptrs came under heavy mortar and gun fire during the induction. The induction was however completed with Cheetah helicopters. The op was a success after the commandos took the post after a 24 hour fight.

May 1999 Op Blizzard 153 HU was called upon to undertake induction of 140 troops in the mercenary infested region of Inshan and Afti area in Delta sector. It was the first such operations where such large numbers of troops were inducted.

1st Jan 2000 Information was received on the area commander of Muzahids along with four major terrorists, hiding in the safe heavens of Hill Kaka area. These terrorists were aiming to enter the Srinagar valley from Pir Panjal pass. Three helicopters were tasked to undertake this challenging mission. They undertook two formation sorties and injected nine SF commandos at the icy height of 9000 ft for a successful engagement.

19th Jan 2000 In the aftermath of infamous Chatsingh Pura incident, where six were brutally massacred, intelligence suggested that terrorists were hiding in the Nawshera area of Awantipura-Srinagar bowl. Reacting to the real time intelligence IAF responded and a three helicopter formation of Mi-17s was detailed to drop commandos from Awantipura. A total of five missions were flown to drop the commandos at different and difficult places. As fallout of this timely action the combing action helped eliminating the ANEs within a short duration of six hours. SHBO missions were flown for induction of troops in the icy heights of Kishtwar and Nawapachi in Delta sector on 24 Jan 2000. A large number of hideouts were busted in this mission in an area dominated by ANEs.

23rd May 2000   Based on real time input in the Awantipur–Sringar axis regarding storage of large cache of arms in an ANE hideout the IAF swiftly responded. A total of five helicopters missions were launched in order to encircle the terrorists and carry out pincer movement to bring the ANEs to their knees. The operation was a great success and large amount of arms and ammunition was seized. Warwan sector is at the northern edge of Delta force AOR. This is a region of ice capped mountain terrain with extensive forests thus making it difficult for ground forces to penetrate. The only way to induct troops into this region was by medium lift helicopters of the IAF. Six sorties were flown on 29th May in the densely forested areas of Inshan and Afti in order to induct Special Forces. The mission was successful as it gave edge to the commandos in retaining speed and surprise. Subsequent combing operations provided the army with control of the high ridges.

Op Sarpvinash (May2003)   The Indian Army's Northern Command conducted a complex terrorist camp-busting operation called Sarp-Vinash. This is one of the landmark counter-terrorism operations in J&K. The Indian Army launched one of its biggest terrorist hunting operations in the Surankot forests from Hill Kaka. It is on the south-eastern slopes of the Pir-Panjal range which separates the Srinagar valley from Jammu. It is also an old Mughal caravan root. The helicopters were called upon to insert troops and provide logistical support for a sustained period of time. Intelligence was received that upto 100 militants were holding ground in and around the Hill-Kaka area in Rajouri sector which provided safe passage to militants inducted from Poonch/ Mendar area to transit into Srinagar valley via the Pir-Panjaal pass. The initial recce with Maj Gen Lidder on board was undertaken by Hora and crew. He also led the first four helicopter SHBO mission (two each from 130 and 153 HU) to Bagla Feature. The hideouts after location were raided by Special Forces and 13 Pakistani militants were killed in the combing operations which lasted for a total of 10 days. Mi-17 and Mi-35 gunships were extensively used to provide suppressive fire and induct troops in the face hostile fire.

Op Sarp Vinash would not have been possible without hard intelligence and tactical airlift. Three Mi-17 helipads were dedicated for logistics, quick relocation of troops and maintaining the integrity of the cordon and stops

a r o u n d Hillkaka. IAF helicopters with an impressive array of direct firing weapons and cannons were used extensively for flushing out the ANEs and p r o v i d i n g suppressive fire. Mi-17s flew induction sorties for troops even by night. Op Sarp Vinash has been instrumental in establishing peace and stability in the entire Rajouri sector; needless to say the success achieved would not have been possible without the contribution of Mi-17s in fulfilling its role. The Mi-17s of 153 HU alone flew more than 20 sorties for these operations. About six sorties were flown in the valley towards clean up by Special Forces.

Op Baraub (July 2005) Most successful CI ops of the year took place in the valley wherein live SHBO mission was flown that saw the army eliminating militants without suffering a single causality. 153 HU got its first VM gallantry medal in this Op.

Op Foxhunt (Feb 2006) On 2nd Feb 2006, Mi-17s were called upon to induct 56 troops of 3 Para SF in Gumbar and Baramulla utilising two helicopters. The formation flew a total of 7 Hrs 30 Min in twelve sorties. This mission in particular proved to be the spearhead for further special operations and gave great advantage to the Special Forces to achieve their objectives.

Aug 2007 On 10th Aug 2007, one Mi-17 inducted 16 troops of 10 RR (Gazali Team) in Doda sector towards intercepting terrorists. Four hours were flown in six sorties to achieve the task. On 22nd Aug, a combined effort of two helicopters was utilized to drop 15 troops in eight sorties. The drop was carried out to induct troops of 17 RR at Chandrakot helipad. On 25th Aug, 22 SF were inserted at Kishtwar by a Mi-17 for a covert counter insurgency operation on basis of hard intelligence inputs.

2010 In 2010, in a routine aerial survey mission a RPA picked up movement of ANE heading towards Amarnath temple. The helicopters were tasked to drop troops to neutralise this threat. The mission was successfully carried without any time delay thereby averting the ambush on the pilgrims.

PROFILES IN COURAGE: CITATIONS

Flight lieutenant Bhupinder Singh Chandel (9726) Flying Branch (Pilot)

On 24th January 1971, he was detailed to evacuate Army personnel from a picket in the Northern Sector. These men were suffering from acute frost bite and were shock sustained during an expected blizzard and avalanche. The visibility was very poor and the post was covered with a thick blanket of snow. While flying over the area, Flight lieutenant Chandel found that hoisting of casualties by the Rescue Hoist was not possible due to strong winds. He, therefore, executed landing on a difficult site and succeeded in evacuating five men, in a serious condition, to the base hospital. In doing so, he displayed a complete disregard for his personal safety, and airmanship of a very high order. It was because of his fine performance that five men were saved from further suffering, and possibly even death.

Flight Lieutenant Abdul Nasir Hanfee (16077), Flying (Pilot)

On 23rd September 1987, he was launched in Siachen area to assist the Army. Taking off in a short time, he immediately undertook the much needed Army support mission in the area. At a time when all our forward posts were under attack by the adversary and the demand for ammunition and other logistics support came up, he rose to the occasion and accepted the challenge. Flying from sunrise to sunset in marginal weather conditions with frequent snow blizzards and white-out conditions and frequent adversary challenging of the posts, he achieved 14 to 15 sorties a day for three continuous days, with total disregard to his personal safety. He singlehandedly managed to provide the much needed logistic support for two days which was a big morale booster for our troops in Siachen Glacier area, fighting the adversary in a hostile terrain. In a restricted area where the guns were required to fire continuously to meet the adversary's challenge, he also showed exceptional professionalism by evacuating 38 casualties in single days from the area with complete disregard to his personal safety.

Squadron Leader Anand Jaivanth Rao (10502) Flying (Pilot)

Sqn Ldr Anand Jaivanth Rao (10502) Flying Pilot was one of the senior pilots of the Cheetah task force, who were assigned to undertake the initial induction of troops, into the Siachen Glacier, during an operational exercise. He was in the first aircraft, to carry out initial recce of the glacier, prior to the commencement of the induction of the task force. On 13th Apr 2004, he carried out induction of troops at an unprepared, unmarked snow bound helipad, located at an elevation of about 17,000 ft. Thereafter, he continued to operate for 20 days at a stretch and inducted troops. He was the first pilot to explore and land at the site of the planned assault camp. With great tenacity for hard work, judgment and high sense of responsibility, he carried out a number of mission in the glacier area, under the most difficult and treacherous flying environments, where minor errors could lead to disaster.

He was also detailed as the leader of the task force to undertake the salvage operation of two Cheetah helicopters from crash site to the assault camp. It was an exacting operation involving underslung flying, which he conducted with the utmost speed and safety. He completed the retrieval of all the aircraft parts, well ahead of planned schedule and undertook 43 sorties for the purpose. Flying for this task was invariably at the limits of performance of the aircraft. During the exercise Meghdoot he flew over in the Siachen Glacier, under all types of weather conditions and displayed outstanding grit, determination and professional competence of an exceptional order.

Squadron Leader Gurmohinder Singh Bajwa (10514) Flying (Pilot)

During the task of induction of troops in the operational exercise "Meghdoot", he carried out a number of sorties in the Siachen glacier, under one of the most hostile environment for flying, with no references for navigation and exacting a landing with snow getting churned up due to rotor wash at the time of landing, he flew his helicopter with great skill and grit to achieve all missions assigned to him. He displayed tremendous endurance for hard work and was always a ready volunteer to undertake additional flying commitments, in spite of being fully aware of the hazardous flying condition prevalent over the glacier. He carried out over ninety missions, lifting armament and troops to various locations on the glacier and continued to operate in the area over extended periods at a stretch with a very high sense of duty.

Squadron Leader Surinder Singh Bains (11288) Flying (Pilot)

Squadron Leader Surinder Singh Bains (11208) Flying Pilot was one of the senior pilots of the task force of Cheetah helicopters, which were deployed for the induction of troops, during operation "Meghdoot", commencing April 84, till completion of the planned first phase six days. He carried out forty sorties involving 42 hours of flying under the most hazardous flying environments. Flying over the unchartered expanse of the Siachen Glacier, the world's longest glacier, located in the midst of towering mountains, he was in the first helicopter to select the landing sites at all strategic passes. These sites were located above 17,000 feet height and at unvisited areas.

It was a display of rare skill and courage of the highest order by him considering the area being devoid of any sort of reference points and completely covered with ice and snow only, with uncertainty of the firmness of the touchdown spot and churning up of snow, by the down wash during landing. Besides, the ambient and in-flight temperatures ranged between – 20º to –30º C most of the time. He was the first pilot to land troops on two different helipads on 13 Apr 84 and 17 Apr 84. During these missions, besides the treacherous flying environment, there was a great deal of uncertainty about hostile activity in the area, as such he had to prepare himself for all eventualities. He not only operated his helicopter with skill and confidence, but also trained other junior pilots in such exacting operations, and thus ensured that the maximum load was lifted in each sortie. Never before in the history of helicopter operations have artillery weapons been carried underslung at such altitudes. This was successfully achieved by him

Squadron Leader Suhas Clement Solomon (10987) Flying (Pilot)

During Oct 84, there was an urgent requirement of air-maintenance at Gyong La 3 post, situated at 4750 meters in the Siachen Glacier. The post has constantly remained under the surveillance of opposing forces. Prevailing severe winter conditions made it imperative that air drops of fibre glass huts and kerosene be carried out on utmost priority. Mi-8 helicopter, while operating in this area is not only fraught with danger of operating beyond its flight envelope, but each mission at Gyong La 3 Post, came under heavy machine gun fire from opposing forces. He was specially chosen for this "High Risk" task. Unmindful of his personal safety, he displayed unflinching courage, determination, dedication and successfully flew 45 missions in 4 days, and carried out precision drops in spite of marginal weather conditions. To avoid coming under Machine gun fire from opposing forces, he had perforce to fly at height as low as 10 meters. His actions not only alleviated the sufferings of the troops and raised their morale, but were highly appreciated by Senior Army and Air Force officers. He has so far flown 211 "High Risk" missions over the Siachen Glacier since Apr 1984, inclusive of 60 missions at Gyong La 3 post. He has airlifted 90.225 tons of vital combat loads and essential supplies. In an outstanding feat in these operations, he created history in helicopter aviation by landing Mi-8 helicopter at a height of 4700 meters with the purpose of evaluating the helipad for landing of expensive and high-demand fibre glass huts, which were suffering damage in spite of low drops.

Squadron Leader Harbans Singh Bath (12008) Flying (Pilot)

He was inducted into Operation Meghdoot in the month of March 1984. He has been actively operating in the Siachen Glacier area and has flown as many as 170 sorties. On many occasions, Sqn Ldr Bath has undertaken air maintenance sorties under inclement weather conditions and within enemy fire to C-3 and Billa Fond La dropping zones and has successfully dropped the delicate and cumbersome loads like pre-fabricated huts without any damage, displaying dauntless courage and precisions. He is one of the pilots to have undertaken trial landing at forward logistic base at the height of 4.7 Km to explore the possibilities of casualty evacuations and landing supplies in extreme emergency, thus creating a world record by landing Mi-8 helicopter at such an altitude. All these tasks involved using the helicopter at its peak performance under adverse conditions which demands a very high degree of competence and professional skill of the pilot.

Flight Lieutenant Anshu Kumar Matta (15875) Flying (Pilot)

Flight Lieutenant Anshu Kumar Matta (15875) Flying (Pilot) was located at Base camp during operation Meghdoot on Siachen Glacier as captain of a helicopter detachment. On 13th Aug 1985 at 0815 hrs, company commander, Captain RK Vasal of 8 Maratha Light Infantry located on an extremely difficult post passed a message to officer-in-charge Cheetah helicopter operation that due to heavy enemy shelling, one junior commissioned officer and one person was seriously wounded and could be saved only if he was speedily evacuated. Flt Lt AK Matta was close by and heard the message being conveyed. He, without wasting a moment for obtaining clearance, immediately took off leaving the clearance for the sortie to be obtained by the Task Force Commander. Not knowing the exact location of the casualty and to avoid delay in reaching the wounded soldier, he landed in the area of Company Headquarters which he was familiar with and picked the Company Commander in his helicopter.

With utter disregard to his personal safety, he reached the area where enemy shelling was still continuing. The place where the wounded NCO lay was at an altitude of 19200 feet above mean sea level. Due to heavy snow, no helipad could be constructed at such a short notice. He manoeuvred and landed his aircraft at that altitude under bad weather and poor visibility conditions. Against all odds and fully aware of the risks, he lifted the casualty and brought the wounded person to Base Camp for first aid. He only waited till the helicopter was refuelled and immediately took off and brought the casualty to 153 General Hospital at Leh.

Flight Lieutenant Ravi C Pathak (23746) F(P)

On 03 Oct 2000 Flt Lt Pathak was to fly as co-pilot in No-1 aircraft of a formation for an maintenance sortie in Op Meghdoot area. Till Camp-III post the sortie was uneventful. After No.1 took off from Kumat, he did not receive any RT call from No.2, which was contrary to the briefing. Realising something was amiss, No.1 Turned back to look for No.2. Flt Lt Pathak performed his duties as a co-pilot admirably and assisted his Captain in locating the position of No.2, which had unfortunately crashed.

There was no safe landing space close to the site. Realising the gravity of the situation, he unstrapped and jumped out of hovering aircraft without giving second thought about the dangers involved. He jumped from 10 feet and walked to the crash site, which was full of boulders. It was an extremely difficult task to walk on terrain, un-acclimatised, at an altitude of 15000 feet, without any oxygen, in sub-zero conditions. He promptly analysed the situation and checked the condition of both the injured pilots. After finding no pulse in Flt Lt VK Murthy, he straightway gave attention to Flt Lt R Sharma and marshalled the army helicopter to a safe place for a low hover. He along with medical Officer put Flt Lt Sharma in to the hovering helicopter. Thereafter he jumped into the hovering aircraft to get some help from the nearest post. After receiving the message that Flt Lt VK Murthy still had some pulse he landed at a safe place near the crash site and dropped captain and medical officer and took off single pilot to Kumar post for help. He analyzed the situation and the gravity of time limitation, which became handy in rescuing Flt Lt Murthy on time.

Squadron Leader Sanjeev Kumar Bhatnagar (18838) F(P)

On 21 Feb 03, he was tasked with a casualty evacuation from Batagul in Drass Sector. On completion of the mission, Commanding Officer of the local army unit requested him to evacuate a patrol of 8 Jawans who were trapped in an avalanche very close to the LoC. He immediately carried out a recce and evacuated all Jawans in two quick and meticulously planned missions in unfavourable weather conditions under the threat of enemy fire. The patrol was stranded in the open without any shelter close to the LoC.

The return route was blocked due to heavy snowfall and avalanches. Without limited supplies and deteriorating weather condition, any delay would have resulted in the loss of precious lives. The rescue was carried out in a calm, prompt and professional manner despite the high altitude (which degraded the helicopter handing and performance), the narrow valley (which made manoeuvring difficult) and the unprepared and soft sloping ground which made a landing difficult). Notwithstanding the perils of deteriorating weather condition, white out conditions and loose snow, which made the task even more hazardous and the omnipresent danger of enemy firing, he gallantly saved the lives of the soldiers. The mission was a display of exemplary courage in the highest traditions of the Indian Air Force. This brave act was well appreciated at all levels by the sister service.

Flight Lieutenant Naresh Bhalerao (22908) F(P)

On 05 Apr 03, he was tasked with casualty evacuation of Capt AK Chandran of 21 Grenadiers who had been engulfed in an avalanche near Chandan II in Op Meghdoot Sector. He reached the site within minutes and realized that the mission would have to be carried out in full view of 'U-cut' a Pakistani observation post, which could direct artillery and mortar shelling on to him. Realising the gravity of the situation, he instructed his No 2 to stay clear while he picked up the casualty by touching down on a single skid on an uneven rough surface and immediately set course for Base Camp. FLt Lt Bhalerao displayed exceptional courage, professionalism and decision taking ability while carrying out the mercy mission at 16000 feet, where the handling and performance of the helicopter is severely affected. The mission was carried out under continuous threat of enemy fire and further avalanches besides the risk of getting into ground resonance and toppling. Notwithstanding the perils, he gallantly evacuated the casualty and ensured timely medical attention. His actions speak of great dedication, courage and unselfishness, which are in keeping with the highest traditions of the Indian Air Force.

OPERATION SAFED SAGAR: ENABLING OP VIJAY

In 1999, the traditional adversary of India, Pakistan, carried out a misadventure of a mammoth scale in the Kargil-Dras sector. Regular army troops of Pakistan came and occupied the peaks of Kargil and Dras in the guise of militants. The infiltration was largely unnoticed until the occupation reached its critical threshold. The Pakistani forces were now present in a position of advantage, overlooking most part of the National Highway 1-A, and directly threatening the Kargil Valley.

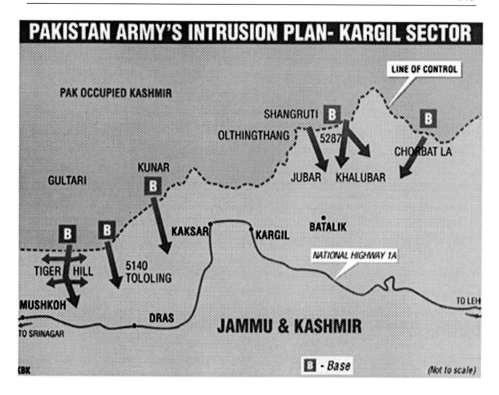

PAKISTAN ARMY'S INTRUSION PLAN- KARGIL SECTOR

The Indian Army swung into action and carried out several assaults incurring heavy casualties due to the tactical disadvantage of having to carry out the operation in full view of the adversary positioned above. It was soon realised that air operations were imperative to flush out these intruders. The Air Force pooled in all resources and initiated Operation-Safed Sagar. As per an Indian Army Headquarters assessment, Pakistan had the following military aims: cut off the strategic National Highway 1A (Srinagar-Leh highway); alter the status of the Line of Control (LoC); give impetus to insurgency in Kashmir Valley and elsewhere in J&K.

The Indian Air Force responded very quickly after the CCS approved employment of air power on our side of the LoC. It deployed its forces and launched the first close-support air strikes with MiGs and armed Mi-17 helicopters within 48 hours. After 23rd May, there were no professional differences whatsoever that could affect our teamwork or planning. After the CCS meeting on 24th May, the three chiefs were closely involved in the politico-military decision-making process. While it would take more than a book to document the IAF's gallant, innovative and successful campaign – Op Safed Sagar, the following narratives highlight some of the contributions of the helicopter fleet.

129 HU: Kargil Diary

The Valley has remained a bone of contention between India and Pakistan ever since independence and the two nations have gone to war over it three times since independence. In 1971, the entire valley of Kashmir had been saved for India by a handful of Dakota crew of 12 Sqn. History repeated itself when Pakistan, violating all international norms, occupied the Kargil Heights.

As per IAF intelligence, more than 600 intruders had violated the LOC and captured 16 posts in Kargil, Drass and Batalik sector. This, however, was an underestimation since the numbers involved turned out to be much larger than expected. The ragged heights of Batalik, Kargil, Drass and Muskhoh valley to the northeast of Kargil, virtually became the eye of the storm. In seven days, the IAF planned its operation Safed Sagar, and Mi-17s from all over started reaching Air Force Station Srinagar as a task force. Three fighter squadrons had also moved in to Srinagar airfield. On 25th May, a joint briefing was conducted in which, after careful inspection of the relevant facts, the fighter fleet projected a 5% success probability. Wg Cdr Sinha who was the Task Force Commander for the Mi-17 fleet took on the challenge.

At 1500 hrs on 25th May, the Chief of the Air Staff was flown to the target area by Sinha and Tyagi. The final decision to go ahead came in the evening. The Indian Army had launched a ground offensive since 12th May; it was our turn to do something. The officers of the unit who took active part in the strike missions were: Wg Cdr AK Sinha; Sqn Ldr PK Mohey; Sqn Ldr NS Verma; Sqn Ldr AS Pande; Flt LT MS Bhandari; Flt LT S Tyagi; Flg Offr S Srinivasan; Flg Offr M Kothari; Flt Lt V Kamthan.

A four helicopter mission was planned for Tiger Hill on 26th May for which a takeoff was planned at 0530 hrs. The technical officers, Flt Lt Jamuar and Fg Offr Konwar of 152 HU, along with air warriors worked whole night on the helicopters. Two were prepared and harmonised in the night with the help of lighting from the headlights of an aircrew transport. Each aircraft was loaded with 128 rockets (RP 57mm). Bullet proof jackets were procured in the night, and the aircrew briefing finally got over at 2300 hrs in the night. Bullet proof helmets were to be brought from the army at BB Cantt and positioned for the crew at the Mi-17 dispersal by 0500 hrs on 26th morning. The stage was set.

On 26th morning Tyagi and Kothari took off pre-dawn at 04:30 hrs for BB Cantt to fetch the helmets and landed back with the load at 05:00 hrs. As the last rocket was being loaded in one of the helicopters, the first rays of the sun appeared on the eastern horizon. The first mission was planned for a strike at tiger hill with four aircraft armed with 128 rockets each and simultaneously two aircraft were planned for a strike at Tololing. The target recognition was done by an army officer on board for each strike, both belonging to the army aviation unit at Sharifabad. Targets were identified and attacked with accuracy. The strike on Tiger Hill was led by Sinha with Srinivasan, followed by Pande and Bhandari in the second helicopter.Dogra and Nitish were in the third helicopter. The strike on Tololing (height of 4.5 km) was led by Kulkarni and followed by RK Singh and Basu. At 1100 hrs, the decision for second strike over Tiger Hill came. Four helicopters were planned. Armour 1 comprised Kulkarni with Muhilan while RK Singh captained Armour 2. Nubra 1 and 2 comprised Sinha with Mohe and Pande with Kothari. The mission delivered the entire fire power with accuracy and landed back at 1200 hrs. At 1230h, a message was received that Flt Lt Nachiketa flying a MIG-27 had been hit and ejection was over Batalik sector. We were still debating on the plans of his retrieval when 20 minutes later another call informed of Sqn Ldr Ahuja flying a Mig-21 FR mission shot down. A Combat SAR mission is debated upon: the Batalik sector is heavily defended by enemy SAMs. About half an hour later the detachment was informed that Nachiketa had been captured and Ahuja shot dead. This was the first taste of war and death that brought home some hard realities.

On 27th, a four helicopter mission was planned for Tololing, each carrying 128 rockets. The crew is Kulkarni with Mohey followed by Tyagi Basu, RK Singh with Nitish and Pundir with Vashisth. The first mission in the morning had to be abandoned due to presence of enemy aircraft over target area. The same formation struck Tololing at 1715 hrs and rockets pounded Point 4590 and the Interior Ridge. Results were encouraging: targets were pulverized, and the army fought a fierce battle at night and captured the Tololing feature.

On 28th morning, a six aircraft mission was planned for the helmet feature at Point 5140 and the Interim Ridge from Tololing. Finally, only four aircraft could take off. The crew composition was: Sinha with Pande followed by Verma with Kochar, Muhilan with Pundir and Nitish with Bhandari. The rear areas of the Mi-17 helipads were under Pakistani artillery fire. No-3 aircraft of this formation, flown by Muhilan and Pundir was hit by a stinger missile and the aircraft came down to mother earth as a fireball. It burnt for the next seven hours. This was the third aircraft shot down by a SAM on the third day of airstrikes. The intruders were well prepared, hard-bred, army regulars of Pakistan. They forced our top brass to re-think; as a result all rocket strikes by Mi-17s as well as fighters were called off, with only high level bombing allowed. The Mi-17s reverted back to their original task of casevac, troop induction and air maintenance. The unit undertook extensive casevac missions from Kagil, Gurmi and Mughalpura. These were occasions when the aircraft landed after sunset while undertaking these missions. The body of late Sqn Ldr A Ahuja was airlifted by our aircraft on 28th May.

Op Safed Sagar continued till first week of July and the unit manned a four aircraft dett at Srinagar for the entire month for June. The bodies of three Pak soldiers were carried by our aircraft on 5th Jun from to Kargil to BB Cantt. Over and above casevac missions, the major thrust remained on air maintenance sorties. Not only were logistics supplies airlifted but vital ammunition and also artillery guns were positioned by our aircraft. These 105mm artillery guns, each weighing 3.4 tons were positioned at a forward post in the Amar sector by our helicopters. Airlift of VVIP/ VIPs to the battle area was also undertaken by the unit, including the airlift of prime minister, home minister and the defence minister. On 13th Jun, the Prime Minister was airlifted to Kargil. This proved a morale booster to all the participants of the operation. The Prime Minister was familiarised with all the strategic posts by the pilots from the cockpit itself while flying over the battle areas. Trials were also undertaken for the feasibility of undertaking night missions.

The Statesman: Mi-17s and the Art of Flying High

MATAYAN, Jun 4 – A Squadron Leader has just completed his "mission', hovering about 300 meters above the bunkers dug by intruders at Dras, firing high-explosive rockets at them. That he is standing in a meadow carpeted with blue and yellow flowers, nonchalantly smoking a cigarette makes his mission even more unreal, for he has seen his comrades die in battle and considers himself lucky to be alive.

"I was flying alongside the Mi-17 gunship that was shot down. I was in the first helicopter, the third one was hit by a Stinger missile," he said. He has flown four missions since the air strikes on infiltrators began on 26th May. Like most Mi-17 pilots he is one of the unlikely heroes of the war. No fortification, from forts designed by Michelangelo to intruder's bunkers high in snow covered mountains, has proved more difficult to destroy. A straightforward infantry assault would cause unacceptable losses: artillery fire can be inaccurate, fighter-bombers flying over tiny targets at Mach 2 can be ineffective, and the much-vaunted Mi-25 attack helicopter cannot function at 14,000 feet. Which bring us to pilots like this man and their Mi-17s, unglamorous but effective transport-copters turned killers. It's almost like Sad Sack becoming a war-hero. The Mi-17 has been fitted with 128 rockets to blast the sangars".

152 HU

In May 1999, the country woke up to the Kargil conflict, and the Mighty Armour (152 HU) were proud to be a part of IAF's Operation Safed Sagar. In fact, the first air strikes over the Tiger Hill and Tololing peaks by the unit on 26th announced the beginning of Op Safed Sagar. It was inducted into the Srinagar valley on 13th May with five helicopters and operated there till 31st Jul. Owing to the urgent nature of the task, the preparation time available was just 13 days, which involved configuring the helicopter for armament role, harmonization of rocket pods and practice of high altitude firing at Toshe Maidan range. We carried out five live offensive missions from 26th to 28th. In the first two days itself, the unit carried out timely and accurate strikes on enemy strongholds which led to the final assault by the ground forces to regain lost territories. The unit also carried out underslung operations of AD guns for the army at LOC pickets from Srinagar and Kargil. 411 sorties clocking 160 hours were achieved and 207 casualties evacuated. On 28th, Sqn Ldr R Pundir, Flt Lt S Muhilan, Sgt PVNR Prasad and Sgt RK Sahu undertook to fly a live Battlefield Air Strike mission at Point 5140 on Tololing Hills as No.3 in a four-helicopter formation. While breaking away, the helicopter was shot down by a Stinger missile. All four of them were awarded Vayu Sena Medal (Gallantry) posthumously. To honour them, a war memorial stands today at AF Station, Sarsawa and various functions are held in their memory. During Op Safed Sagar, the unit flew 59 sorties and 37:15 hours.

128 HU

In the initial days of the operation, a Mi-17 helicopter of 152 HU Sarsawa was lost to enemy fire. Flt Lt Muhilan, who had been a Siachen Tiger, gave the supreme sacrifice and laid down his life in the icy heights of the Dras sector. At the same time, two pilots from 128 HU were attached to Sarsawa to augment the efforts towards Op-Safed Sagar. With tensions increasing between India and Pakistan, the rest of the Tigers were kept on 'stand by' to move to the Northern sector. All were more than keen to move into operations. Air maintenance was stopped and helicopters were fitted with rocket pods. During the same time, Ahluwalia and Verma had moved to Srinagar and were flying extensively in the support of the Army in the entire area spanning from the Srinagar up to Leh. The sorties were primarily towards SAR and casualty evacuation. In addition to this, massive air effort was put in to deliver supplies including ammunition to the soldiers deployed on the war front. On 9th June information was received that four more pilots had to go to Srinagar for Kargil operations. This was followed by a rush to the Flight Commander's office by all unit pilots volunteering for the task. Finally, on 11th June, four lucky pilots, Robin, Kathuria, Manish and Dobhal were launched to Guwahati from where they were air lifted to Srinagar to fly in Op-Safed Sagar. The CO, Wg Cdr Isser joined up a few days later. They operated from Srinagar and flew 62 missions braving enemy fire and going up to the far reaches of the theatre of operations. A total of 224 battle casualties were air lifted, some under extremely hostile conditions. The airlift of four 130mm artillery guns to Chakwali in Gurej Sector, in a single day and under enemy shelling, was a feather in the Tiger's cap. Four sorties of this mission were operations beyond the envelope of the Mi-17, but in war it is only fair to innovate! This mission was led by the CO himself. He also led numerous missions with Special Forces (6 Para), including to Bakarwal under enemy fire.

132FAC Flt

The flight was tasked to carry out various missions like Air Borne Forward Air Control, Mobile Observation Posts deployment, reconnaissance, communication duties and SAR operations. It maintained a four aircraft detachment at Srinagar from May 1999 to 19th Jul. Its expertise was tested to the limit when it undertook a total of 50 Air Borne FAC missions; the highest being in the month of June when it undertook 24 missions. During the entire period, 252 hours in 395 sorties undertaking variety of missions and operating from an altitude of 5000 ft to 17,000 ft in vast sectors was executed. During the same period, the flight also undertook 48 casevac sorties saving 82 souls. For displaying true professionalism in the time of need, five officers of this flight were awarded gallantry awards for their dedication to duty during Op Safed Sagar. The Kargil scenario ended in July 1999. It was decided to air maintain various far flung posts in sub sector west in the area of operations of 8 Mtn Div. The flight was tasked to maintain a three helicopter detachment at Khalsi in order to undertake air maintenance task. It undertook air maintenance operations Ex-Khalsi for 'A' & 'B', Jhunkar valley from Oct 1999 and continued the operations till 10th Jan 2002. During the entire period of operation, the flight flew a total of 1009 hrs in 2655 sorties to achieve a total tonnage of 501.53 tons. The highest monthly tonnage was in the month of Nov 1999 when it lifted 66 tons in 135 sorties and flew 110 hours towards air maintenance. Along with the air maintenance, the flight also augmented army efforts in undertaking army casevac and was responsible for saving 157 souls.

TRIBUTE TO THE GALLANT: AWARD CITATIONS

Vir Chakra

Wg Cdr Anil Kumar Sinha (16074-T) F(P)

On 28 May 1999 Wg Cdr Sinha was leading a helicopter formation tasked with aerial attack on point 5140 of Tololing ridge using air to ground rockets. Enemy forces stationed on this side were carrying out accurate firing on the national highway connecting Srinagar with Leh. The enemy was equipped with ground to air shoulder fired stinger missiles and were firing the missiles at the formation. Two of these missiles barely missed Wg Cdr Sinha's helicopter and it could be seen on the video recording. During this attack, one of the missiles hit the number three helicopter in the formation and all its four crew members were killed. The action of the helicopter formation was effective and the enemy suffered heavily at their hands. Wg Cdr Sinha's display of courage in facing grave and eminent threat to himself and the helicopter, and the leadership and devotion displayed by him is in true spirit of the rich tradition of the Indian Air Force.

Vayu Sena medal (Gallantry)

Squadron Leader Ajai Prakash Srivastava (17471) F(P)

The task of Airborne FAC for fighter strikes was assigned to the unit. To overcome the problem of pin pointing enemy target position on ground, he worked out an innovative idea of lighting up the target with colored smoke from Arty fire and also carrying an Air Op pilot on board to direct this fire over targets. Missions undertaken incorporating the above were found to be very successful. Sqn Ldr Srivastava himself undertook numerous FAC missions in Dras, Batalik, Mashkoh areas and was instrumental in guiding fighter strikes over Tololing and Tiger hill in particular. These missions required extensive and thorough planning on his part in not only avoiding heavy enemy small arms and rocket fire but also staying for as little time as possible in enemy missile range. Along with this he also undertook the training of other pilots so as to enable them to undertake these missions independently in tactical battle area. He also undertook numerous sorties towards deployment of observation posts on LOC. In these demanding circumstances he succeeded in keeping the entire detachment at its peak and brought out the best in them.

Squadron Leader Narender Singh Verma (18298) F(P)

On 28th May 99, Sqn Ldr NS Verma was detailed to fly as Captain of No 2 helicopter in a four helicopter formation on a live air to ground rocket firing mission. The target, Point 5140 was located on the Tololing Ridge in Dras sub sector at 16,500 feet. Enemy positions to be attacked were entrenched in well concealed bunkers, thereby making target acquisition difficult. The target area was heavily defended as the enemy had deployed stinger missiles and 37 mm anti aircraft guns in the sector. These weapons posed a grave threat to the slow moving helicopter which was as such operating at the extreme limits of its performance envelop. For the air strikes to be effective, he had to fly close to the target whereby it came well within the effective range of the Stinger missiles. During the attack, Sqn Ldr NS Verma's aircraft was fired upon by missiles. Despite enemy fire Sqn Ldr NS Verma remained undeterred and in the best traditions of the IAF displayed exceptional courage by pressing on with the attack, thereby successfully achieving his mission.

Squadron Leader Nitish Kumar (19898) F(P)

During Op safed Sagar, the helicopter were tasked for strike mission on enemy strongholds in Kargil & Dras Sectors, Sqn Ldr Nitish Kumar was the Captain of four strike missions which targeted Tololing and Tiger Hill on 26,27 and 28 May 99. The attack was against heavy odds of terrain, altitude and hostile fire (including Stinger Missile and small arms). Handling the aircraft close to the limits of its flight envelope called for not only superior flying skill but also exceptional courage in view of the prevalent hostile environment. The attack was carried out fearlessly and with precision resulting in considerable damage to the intruders. The success of the attack is also attributed to meticulous planning and preparation carried out prior to the mission. Non availability of an ejection seat or force landing field implied that any hit on the helicopter would be critical. Unflinching in the face of risk, he displayed immense courage and fighting spirit and undertook the task with professionalism. His enthusiasm and valour was a major contributor towards boosting the morale of the other crew in the team.

Flight Lieutenant Subramaniam Muhilan (22739-A) F(P) (Posthumous)

On 28 May 99, Flt Lt Muhilan was tasked to fly a live strike against enemy position at point 5140, located at an inhospitable altitude of 5.1 Kms. While undertaking the missions, Flt Lt Muhilan was fully aware that the area was infested with missiles and that the only counter measure against the missile was not fitted on his helicopter. Realising the importance of the mission, he boldly accepted the challenge and with total disregard to personal safety went ahead with the strike. Having successfully pressed home the attack, during the break away, Flt Lt Muhilan's helicopter was hit by a stinger missile. Despite Flt Lt Muhilan's best efforts to control the damaged helicopter it crashed, killing Flt Lt Muhilan and all his crew members. Displaying a very high level of professional competence and courage, Flt Lt Muhilan had already undertaken three strike missions in the treacherous terrain of the Dras sub sector. These missions achieved astounding success and were the key to subsequent easy recapture by the ground forces. In his action, Flt Lt Muhilan displayed exceptional courage and acted in keeping with the best traditions of the Indian Air Force.

MARTYRS OF THE 'NINETIES'

Flt Lt Sandeep Jain 19534 F(P) 130 HU KIA 26 Aug 96

Plt Offr Vaibhav Bhagawat 23456 F(P) 130 HU KIA 26 Aug 96

Sgt R Murigappa 679914 130 HU KIA 26 Aug 96

Sgt Krishna C Jha 678079 130 HU KIA 26 Aug 96

Flt Lt Rajiv Juyal 21727 F(P) 114 HU KIA 02 Jul 1997

Flt Lt Sandeep Jain was flying a supply mission to Hoshiarpur post in the Siachen Glacier area, when his Mi-17 Helicopter was shot down by hostile fire.

Plt Offr Bhagwat was the co-pilot to Flt lt Jain. The Mi-17 helicopter on the supply mission was lost to hostile fire over the Siachen Glacier.

Sgt Murigappa was the flight gunner on board the Mi-17 shot down by hostile fire at the Hoshiarpur post in the Siachen Glacier area.

Sgt Krishna Jha was the flight engineer on the Mi-17 Helicopter downed by hostile fire on the Siachen Glacier during a supply mission.

Flt Lt Rajiv Juyal's Cheetah helicopter was lost to enemy ground fire in the Siachen Glacier area during a forward supply mission.

  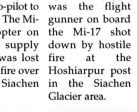  

Flt Lt V K Aggarwal 22185 F(P) 114 HU KIA 02 Jul 1997

Flt Lt Subramaniam Muhilan 22739 F(P) 152 HU KIA 28 May 1999

Sqn Ldr Rajiv Pundir 17413 F(P) 152 HU KIA 28 May 1999

Flt Sgt P V N R Prasad 695490 152 HU KIA 28 May 1999

Sgt Raj Kishore Sahu 729917 152 HU KIA 28 May 1999

Flt Lt Aggarwal was the second pilot in the Cheetah helicopter that was lost to enemy ground fire in the Siachen Glacier area during a forward supply mission.

Flt Lt Muhilan was the pilot of a Mi-17 that was shot down by Shoulder fired missile at Tololing range in the Kargil area during a raid on Infilitrators positions. The crew of four in the helicopter were lost.

Sqn Ldr Pundir was killed in action while flying as a co-pilot of a Mi-17that was shot down at Tololing. They were flying as No.4 in a formation of four choppers attacking enemy positions.

Flt Sgt Prasad was the Flight Gunner on board the Mi-17 helicopter flown by Flt Lt Muhilan. He was killed when the helicopter was downed by shoulder fired SAMs.

Sgt Sahu was the Flight Engineer on board the Mi-17 helicopter flown by Flt Lt Muhilan which was shot down over Tololing by SAM missiles.

Courtesy: IAF website

CHAPTER VIII

The Northeast: Selfless Service to Nation-Building

Introduction   In a TV advertisement for Assam Tourism, a little-known fact springs up – that the Ahom Dynasty in Assam was the only one that no one could subdue, not even the great Mughals. Those of us who have served in the northeast or the 'seven sisters' will acknowledge the indomitable spirit and resilience of the people. The strong sense of identity and other causes have thrown up a series of insurgency movements across all these states since independence. The Indian armed forces have been involved in putting down or controlling these fires through the decades, starting from the Naga problem. IAF helicopters have played a crucial part in supporting the Indian Army in counter-insurgency strategies in Nagaland, Mizoram (the most successful campaign), Tripura, Assam and even Meghalaya and Arunachal. The biggest issue in countering insurgency is the inevitable lack of 'boots-on-ground'. This is where the medium lift helicopters of the IAF pitched in over the decades, mobilizing boots from one place to another; in short acting as enablers and force multipliers. One such story of a firm action in Mizoram is documented in this Chapter. Where required, IAF helicopters have been used in the offensive support role such as against the ULFA across in Bhutan. Counter-insurgency is only a part of the strategy to integrate the north east. The other more important part is holistic development and nation-building activities. IAF helicopters have played an equally important role in that too.

The Border Roads Organisation is doing yeomen service in building up infrastructure all across the seven states. Helicopters play a part in reconnaissance, conveying of workers and officers, airlifting supplies that cannot reach there by any other means, and in many other ways. A partnership between the BRO and IAF station at Mohanbari pioneered a revolutionary concept in 1999. The objective was to trek by ground (aircrew included) along the proposed road axis and select suitable areas as future helipads. Some of these were never-ventured into areas. Thereafter,

Special Operation Mission: Over the Mighty Brahmputra

helicopter would drop or winch in equipment so that a rough clearing could be made. A basic clearing allowed Mi-17s to under slung bulldozers (in manageable pieces) since landing was impossible. This included cumbersome gantries that allowed the heavy equipment to be assembled, and also fuel and rations. The bulldozer would then carve out a helipad, and regular supply of equipment, rations and personnel could begin.

Theoretically speaking, each such helipad would allow two additional road-building avenues. Thus three such helipads on the Chaklagaon axis in Arunachal cut down the road-building time to one-sixth. Such sorties are extremely demanding and on the fringes of the performance envelope of the helicopters. Similar activities have been supported by IAF helicopters in building dams, bridges, hydel-stations, advance landing grounds and many others.

The most visible role, and also the most satisfying for pilots, is saving lives and humanitarian assistance. An unknown fact is that IAF helicopters all over the NE are always on standby to assist civilians in medical evacuation as well as to deal with endemics. Disasters such as the Arunachal flash floods in 2000 and the Sikkim earth quake in 2012 have showcased IAF helicopters giving their professional best in support of their citizens. Thousands of medical evacuations all over the north east over the last four decades are a legacy that is part of the story of integrating and developing the 'seven sisters'.

110 HU: Mizo Operations

The Unit formally got a Commanding Officer on 18th Aug 1963 with Sqn Ldr NK Gaikwad taking over the reins. The contribution to the North East continued and is a saga of innumerable operations ever since. It participated in 'Op Orchid' in 1964 and also in '1965 Ops' where 110 HU provided widespread logistic support to Indian Army in the eastern theatre.

The Mizo problem that had been simmering for a while came to a head on the last day of Feb 1966 when the Mizo National Front (MNF) captured the Aizwal treasury and surrounded the HQ 1 Assam Rifles. Other elements of the MNF also surrounded the Assam Rifles posts at Champai, Darngaon, Vaphai, Lungleh and Demagiri. A detachment of six Mi-4 helicopters of 110 Helicopter Unit, based at Tezpur was sent to Kumbhirgram airfield on 2nd Mar 1966. A sub-detachment of two helicopters was established at Aizawl. The Unit undertook quick reaction operations, including an abortive attempt to raid an MNF camp in East Pakistan near the Mizoram border. It undertook many such missions with 5 Para, 13 Kumaon, 8 Sikh, 18 Punjab, 1 Assam Rifles, 6 Assam Rifles, 16 Jat, 5 Jak Rifles and some other battalions. Successful operations and the picturesque location of 'Kumbhi' surrounded by tea gardens, was an incentive for the men and machines to move in permanently in 1968.

Operations   The CO was told that he would have to request the army battalion (18 Punjab, commanded by Lt Col Raghubir Singh) for accommodation and guards for the helicopters. A seventh helicopter joined the detachment on 3rd March after completing an air test. On 4th March attempts were made to fly in elements of 18 Assam Rifles (AR), which had been moved to Kumbhirgram from Dimapur, into the besieged post of 1 AR in Aizawl. However, MNF elements had occupied vantage points to the North and South of the post and opened fire at the helicopters ferrying in the troops. Majors Sidhu and Balwant Singh, who were in the post, fired a red cartridge to indicate that it was not safe for the helicopters to attempt a landing. All seven helicopters then returned to Kumbhirgram to await further instructions. One had suffered minor damage after taking a bullet hit on the tail boom. It is not widely known that Lt Gen SHFJ Manekshaw, GOC in C Eastern Command, and Air Vice Marshal YV Malse, AOC in C Eastern Air Command, flew in a Caribou aircraft of 33 Squadron for a reconnaissance over Aizawl. The Caribou limped back to Kumbhirgram airfield riddled with bullet holes. One bullet narrowly missed the future Field Marshal who, it is understood, was standing behind the co-pilot during the reconnaissance.

After consultation with Army and Air HQ it was decided to fly in troops into the 1 Assam Rifles camp with fighter escorts. Accordingly, seven helicopters and four French built Ouragon fighters, nicknamed Toofanis in the IAF, were used for this operation. Rendezvous was in the Turial Valley east of Aizawl. As each helicopter turned onto the final approach to the makeshift helipad at the AR Post, one Toofani on each side of the chopper fired rockets at the MNF elements sitting on the North and South of the post. Suffering casualties, the MNF cadres fled the scene and the siege of the post was thus ended. Additional battalions, 2/11 GR, 8 Sikh and 5 Para reached Aizwal by road from Silchar. While Toofanis operating from Kumbhirgram and Hunters operating from Jorhat were subsequently used over Champai, Darangoan, Vaphai and Demagiri; these operations were not coordinated with any helicopter activity and were undertaken to keep the MNF at bay and to ease the pressure off the surrounded posts till they could be reinforced by flying in troops by helicopter.

The new helicopter stream of the IAF did not have experienced pilots in its ranks. Even their COs did not have operational experience. Therefore, the bulk of the flying was carried out by inexperienced and raw Pilot Officers and Flying Officers with a sprinkling of Flight Lieutenants. For the Mizo operations, 110 Helicopter Unit established a detachment at Kumbhirgram and a sub-detachment of two helicopters at Aizawl. Communications between the base and the two detachments was minimal as communications (hand cranked field telephones) were very poor and for all practical purposes could be considered nonexistent. In effect the Aizwal detachment commander, often a Flying Officer or a Pilot Officer had to make his own decisions regarding the feasibility of any sortie being proposed by the Commander 61 Mtn Bde at Aizawl. Despite the absence of experienced pilots, the Mizo operations are a success story as far as helicopter operations go. The young helicopter pilots enjoyed a great degree of freedom with which they could operate, and soon matured into experienced helicopter pilots, capable of undertaking diverse roles in a hostile weather and terrain

environment. They were not averse to occasionally bending a rule or two, but in hindsight this license contributed significantly to the build up of experience which could later be exploited for operations in other theatres.

5 Para, one of the first army battalions to move into the Mizo Hills, was based at Aizawl and

was used as a fast reaction unit of 61 Mtn Bde. While the battalion may have had some role in and around Aizawl, it was primarily used to intercept MNF cadres on the move. Many such operations were undertaken by the battalion in Mi 4 helicopters of 110 Helicopter Unit. It was here that the term Special Heliborne Operations (SHBO) entered the lexicon of joint Army-Air operations of the Indian Armed Forces. Helicopters offered high mobility to troops, and a tremendous element of surprise. An enemy that had been sitting unchallenged for days or weeks could suddenly, without warning, find itself under assault from troops brought in by helicopter. What needs to be emphasised here is that none of the young pilots had been trained for the role. They had just read about the exploits of similar air mobility operations then ongoing in Vietnam and were willing to undertake such missions, fully confident (overconfident?) of their ability and the helicopter's capability to undertake these operations. However, apprehensive that higher Headquarters, with no helicopter experienced officers manning any desk may look upon such initiatives with disquiet – these young pilots downplayed their role in these operations. A comparison of the awards won by personnel of 5 Para with those by the chopper pilots (none), who were equal partners in the operations in Mizoram, would indicate how self-effacing the chopper pilots were about their role.

110 Helicopter Unit undertook many such missions even after 5 Para was de-inducted from Mizoram. These operations were undertaken with 13 Kumaon, 8 Sikh, 18 Punjab, 1 Assam Rifles, 6 Assam Rifles, 16 Jat, 5 JAK Rifles and some other battalions. Lt Gen Sagat Singh who was then the GOC 101 Communication Zone and oversaw operations in Mizoram must surely have come to believe in helicopter operations after his experience with 110 HU. The unit undertook many quick reaction operations at his instance, including an abortive attempt to raid an MNF camp in East Pakistan near the Mizoram border. Gen Sagat's experience with 110 HU perhaps contributed to his confidence in planning for the heli-lifts during the Bangladesh war in 1971 when he was the GOC 4 Corps. The unit played a key role in those heli-lift operations.

TROOPS SLITHERING - INDO-US EXERCISE AT CIJWS JUN 2004

Mi-8 Operations: Northeast

The Mi-4 helicopter was flown for 18 years by 110 HU in various operations within the country and as well in neighbouring countries for different missions. In 1981 this unit was equipped with Mi-8 helicopters. The basic role assigned to the unit was air maintenance for Indian Army and paramilitary forces in Nagaland, Mizoram, Tripura and Manipur. In 1981, equipment and stores of National Hydel Project Corporation were airlifted to Bhairabi in Mizoram. In the month of Dec 1981, during emergency in Mizoram the unit participated in Op JIM-JAM, 368 sorties were carried out towards airdrop, reconnaissance, communication, troops induction and casualty evacuation for military and paramilitary forces. In 1988, 110 HU sent its aircraft to Sri Lanka in Op Pawan, where it flew extensively for various missions in support of IPKF. In the very next year, in 1989-90, it undertook flood relief operations in Bangladesh, Tripura and Assam.

403 Air Force Station was upgraded to 22 Wing, AF and the station celebrated its Silver Jubilee in 1991. In the same year the unit was involved once again in relief operation when a cyclone had hit Bangladesh. In 1992, it actively involved itself in operation Hammer with the Army and Paramilitary forces. Four helicopters took part in the operation from a disused airfield in Palel. In the early nineties, three helicopters were involved in Op Suraj, Op Malar and Op Rajat from Chakabama. In 1995, it got involved in Op Golden Bird with four helicopters. Also, in the same year, it undertook landslides and flood relief operation in Mizoram. From 23rd to 26th Jul 1996, it carried out flood relief operations in West Bengal flying 124 sorties and 104 hours to drop 107 tonnes of load. In 1996, it took part in Op Vajra. It has ever since been actively participating in the CI operations in the northeast. Notable amongst them were Op Ashok in Apr 1997 in Meghalaya and Op Tiger Leech in Mar 1998 in the Manipur and Mizoram sector. The unit also flew for flood relief operations in Purnia in Sep 98.

109 HU   In the 1986 incident of Chinese intrusion in the North Eastern parts of the country, the unit pioneered Op Falcon. Two helicopters were launched to Tawang in Sep 1986. Wg Cdr V Natrajan SC VM commenced operations along the border with China in support of the Indian Army. Under difficult conditions of weather and terrain, the operations were streamlined and given the necessary impetus to sustain the Indian Army now deployed along the Chinese border through the long hard winter ahead. Though the unit withdrew from the scene after two months, the initial spade work done and infrastructure set up by the unit stood the subsequent detachments in good stead. Subsequently, the newer Mi-17s of 127 and 128 HU took over the mantle of enabling a good riposte by the Indian Army to the Chinese incursion.

Mi-17 Operations in NE: 128 HU

128 HU or Siachen Tigers was raised on 30th Dec 1985 at Hindon and became

one of the frontline units of IAF to be equipped with the formidable medium lift helicopter – the Mi-17. The Tigers, soon after their formation, headed for greater heights of the Siachen Glacier across 'Khardung-La' embarking on an enterprise never explored before by medium lift helicopters. Man and machine were pitted against once of the most treacherous mountainous regions in the world with bone-chilling Arctic temperatures. They bagged the trophy for 'Best Helicopter Unit of IAF' in the very first year of their formation and put together a long list of honours and awards. In Mar 1988, it moved to a new locale – the North East. The initial home was at Chabua before moving to Mohanbari in May 1992. The terrain in the east varies from the lofty peaks of Tawang to the humid and lush green forests of Siang, Subansiri, Lohit and Tirap valleys and the challenging Nagaland sector. A heady combination of demanding terrain, unpredictable weather and vigorous tasking of the unit won it a place in helicopter folklore. The peacetime role of the unit is air logistic support for the Indian Army, BRTF and the Government of Arunachal Pradesh. Other important tasks include casualty evacuation, communication and aid to civil power during natural calamities.

Move to Chabua   The unit flew extensively in Op-Meghdoot, scaling new frontiers and setting examples for others to emulate. Soon it was time to move on and relocate to a new area of operation. From the lofty Himalayas of the north, the unit moved to the lush green valleys of the east. The initial deployment was at Chabua. Adaptability being one of the core competencies of the Siachen Tigers, they took little time to get into the grooves of the new of operating conditions of the East. It maintained four detachments from the heights of Tawang to the more amiable, close to sea level detachments of Hayuliang, Along and Pasighat. The unit had barely settled itself in the new area of operations when relief missions were launched. In August 1988 after 10 days of incessant rains, the Brahmaputra overflowed marooning more than 1,00,000 people. A commendable job was done in rescuing men and material from the fury of the river. The unit worked overnight, planning and preparing helicopters for relief operation and air lifting over 60,000 kg of food and supplies to the marooned. Over 35,000 kg of food, medicines and other essentials were air dropped to the flood hit areas of Belurghat totally cut off from the rest of the world. This was the start of a long duel with the mighty Brahmaputra.

Op-Bajrang   Winters are traditionally a picnic time at Chabua with clear blue Brahmaputra waters, sand beaches, and drink in hand with piping hot snacks making for many a cosy afternoon. But in 1990, the political situation and the rising militancy in the state precluded all that. In a swift move, President's rule was slapped on the state and Op-Bajrang was declared with the Army moving into the valley from all over the nation. As a prelude to Op-Bajrang, special heliborne operations (SHBO) were carried out to train 1032 troops by day and night. On the night of 14th Dec 1990, the unit was informed about a heliborne mission to a selected place and were required to carry troops and assault boats. Meticulous planning, thorough briefing and impeccable professionalism were the hallmarks of this joint operation. After briefing troops and discussing the contingency plans, one assault boat along with troops was dropped at the target location on time. Subsequently during the day, 13 missions were carried out. 379 troops and 985 Kg of load were airlifted into hostile and densely forested area for a very successful engagement. This was the start of a long line of involvement and joint counter-insurgency missions for the next two decades.

Relief Operations in Bangladesh   In the beginning of March 1991, with the pre-monsoon clamping down hard, cancelled flights, leaking roofs, flooded dispersals and mundane affairs like weeklong power cuts had everyone gearing up for the rainy months ahead. Towards the end of April, a small low pressure trough in the Bay of Bengal had appeared. It initially appeared to be a run-of-the-mill low pressure, but in the days to follow it intensified into a depression and further on became a mammoth cyclone which veered and hit the Bangladesh coast with all its fury. On the night of 29th April 91, Bangladesh was ravaged by this vicious cyclone, which caused extensive damage to both life and property. The entire chain of the southern islands and regions were inundated by a watery shroud for over twelve hours and hurricane winds of the order of 150 kmph lashed the country. The damage was enormous, with a large number of people killed and rendered homeless. All communications were severed and epidemics broke out.

Bangladesh cried out for help and international relief from UK, USA, Japan, Pakistan, China, Thailand, UAE and India poured in. Government of India sent three Mi-17 helicopters of IAF on 9th May 1991. Siachen Tigers were selected as the Indian ambassadors of goodwill and angels of mercy. A professional team of eight officers and thirty airmen were sent to Bangladesh with three helicopters under the CO, BS Siwach. The team braved unprecedented challenges posed by inclement weather, terrain and alien operating conditions and airlifted over 450 tonnes of relief supplies within a fortnight. During the 15 days of operation, it overcame several hurdles and did an excellent job, leaving behind a fine impression of the IAF with the world community. The Bangladesh Government, as a token of their gratitude, announced the award of Cyclone Relief Medal to the personnel of the unit for their valuable contribution towards relief work.

Op-Rhino Assam continued to go in and out of militant and anti-national activity in the beginning of the Nineties. The government initiated Operation-Rhino to curb the activities of various anti-national forces active in this region. During Sep 1991, the unit was tasked with a mission to drop army troops at a forward location by night due to tactical considerations. The call was answered with an immediate activation and three aircraft got airborne and positioned at Tezu airfield an hour after sunset in total darkness without any landing aids. Most of the evening on the day of the mission was spent sitting over maps, planning tactical routing and carrying out vital calculations to ensure total success of the mission. After a thorough briefing, the helicopters took off as planned. Troops were dropped with pinpoint accuracy on the designated spot exactly on time. Many such successful missions towards counter-insurgency against the ULFA have been done by helicopters of Mohanbari, Kumbhigram and Gauhati.

Move to Mohanbari On 18th May 92 the unit moved to Air Force station Mohanbari. History is testimony to the role played by the small hamlet of Mohanbari in World War II. The renowned 'Hump Operations' were carried out with this airfield as one of the bases. It was thus, a moment of pride for the Tigers to have finally reached a place they deserved to be – a place which in the coming years would give them their rightful place in IAF history. Every installation of the Air Force Station was a reminder of the glorious, action-packed past of this place. Be it the 'Igloo Hangar', the 'old ATC building' or the sight of the Raj-era tea estate bungalows; every moment spent at Mohanbari was reminder of a glorious past.

Army Rescue The unit faced a daunting challenge in the month of July 1997. It was to search and rescue an army Chetak helicopter which went down at Sakra Bari. The aircraft had, apart from the crew, the Chief of Staff of 3 Corps on board. The treacherous terrain and inclement weather made the effort extremely challenging. The wreckage was located after a Herculean effort put in by the Tigers for three days and the bodies of the deceased officers was flown back by the same crew to Rangapahar. The Indian Army maintains posts in practically inaccessible areas in the far reaches of the nation's boundary in the east. On 28th Nov 1998, a call was received from for an urgent casualty evacuation from Tawang. The notice was short and warranted an immediate activation. The SAR helicopter was prepared swiftly and the mission was successful with a landing back by night. Such missions in support of the Indian Army were and are a matter of routine.

A Mi-17 helicopter is an ideal platform for such missions due to its versatility and robustness as well as its capability to fly in adverse weather. On 11th Sep 1999, a call for an immediate casualty evacuation was received from a forward army post in the vicinity of Tawang. Atri and Deshmukh undertook the mission and recorded another success in providing succour to our soldiers.

Flood Relief – May 1998   Somewhere in mid May, one Cumulo-Nimbus cloud got multiplied into a hundred and went unnoticed by everyone. It covered the entire Assam and Arunachal and rained so hard that both the states were inundated. The timely action of the civil administration and HQ Eastern Air Command helped in saving thousands of lives held ransom to nature's wrath. UK and Maneesh were launched at short notice for the massive flood relief operations. In true spirit of the Tigers, the crew saved many a life and brought smiles to the faces of many more. These operations continued till the month of July with several other crews being tasked with the relief missions in different regions of the two affected states. It is not for nothing that the local people in the far flung villages of Assam and Arunachal swear by the name of the Siachen Tigers.

Op-Safed Sagar   At the beginning of the operations, two pilots from 128 HU were attached to Sarsawa to augment the efforts towards Op-Safed Sagar. With the tensions increasing between Indian and Pakistan, the rest of the Tigers were kept on 'stand by' to move to the northern sector. All aircrew were more than keen to move into operations. Air Maintenance was stopped and helicopters were fitted with rocket pods. During the same time, Ahluwalia and Verma had moved to Srinagar and were flying extensively in the support of the Army in the entire area spanning from the Srinagar sector up to Leh. The sorties were primarily towards SAR and 'casualty evacuation'. In addition to this, massive air effort was put in to deliver supplies including ammunition to the soldiers deployed on the war front. On 9th June information was received that four more Tigers had to go to Srinagar for Kargil operations. This was followed by a rush to the Flight Commander's office by all unit pilots volunteering for the task. Finally, on 11th June, four lucky pilots, Robin, Kathuria, Manish and Dobhal were launched to Guwahati from where they were air lifted to Srinagar to fly in Op-Safed Sagar. The CO, Wg Cdr Isser joined up a week later. The Tigers operated from Srinagar and flew a total of 62 missions braving enemy fire and going up to the far reaches in the theatre of operations. 224 battle casualties were air lifted, some under extremely hostile conditions. The airlift of four 130mm artillery guns to Chakwali in Gurej Sector, in a single day and under enemy shelling, was a feather in the Tigers' cap. This mission, four sorties of which were beyond the operating envelope of the Mi-17, was led by the CO himself. He also led numerous missions with Special Forces, including to Bakarwal and Kaobol Gali under enemy fire.

Flood Relief June 2000   June was a maddening monsoon month in the year 2000. It was raining the entire month and on 12th June tragedy in the form of flash floods struck Siang district of Arunachal Pradesh. Due to sudden rains the water level in the Siang River rose by 10-15 metres in just an hour. A few unlucky souls never got a chance; but to save the rest the Tigers rushed in. They flew from sunrise to sunset and the unit carried operations from Pasighat, Yinkiong and Along to airlift 2,389 persons including winching up of 265 critical life and death cases. A total of 48 hours were flown in 162 challenging missions to deliver 36,720 kg of emergency ration and medicines. At midnight of 12th June, news was received of a steamer stuck in the swirling waters of the Brahmaputra. In the dramatic rescue, 98 passengers were recovered safely from the sinking boat. The passengers were flown to Passighat but on reaching there it was found that the ALG was flooded. The CO took the split second decision to divert to Mohanbari. In recognition of this effort, three unit personnel-Isser, A Sharma and Sgt SK Singh were awarded the 'Vayu Sena Medal'.

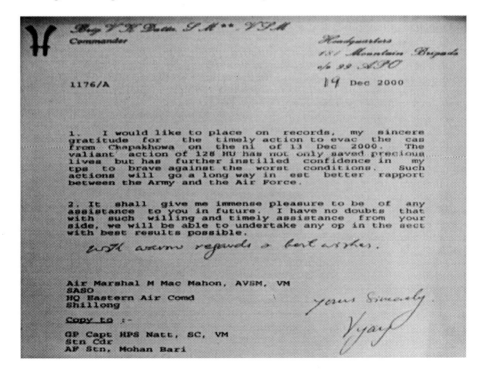

Army Cas Evac – Chapakhowa, Dec 2000   In Dec 2000, insurgency in Assam was at its infamous peaks. The Indian Army was deeply involved in operations to counter the miscreants and their anti-national activities. During one such operation, a unified command party on a patrol in the foothills was ambushed. Four soldiers were critically injured in the ensuing gun battle. A request for an immediate casevac reached the unit. It was beyond sunset time when the news was received and the unit immediately pressed into action. A night casualty evacuation mission poses definite challenges to both the flying and the ground crew. The details provided about the exact location of the troops was inadequate and this coupled with the hostile environment prevailing at the site made the task even more intimidating. The SAR aircraft was prepared in record time by the ground crew even under low light conditions. The mission was undertaken by the Commanding Officer Wg Cdr R Isser along with Sqn Ldr NK Atri. A thorough and swift preparation on ground ensured smooth conduct of the mission. The injured soldiers were picked up from the Chapakhowa village and transferred to MH Dinjan by 2330 hrs. This was done at a time when night vision goggles had not come in to the country. Accolades poured in from the Raksha Mantri, the Army Chief and many others

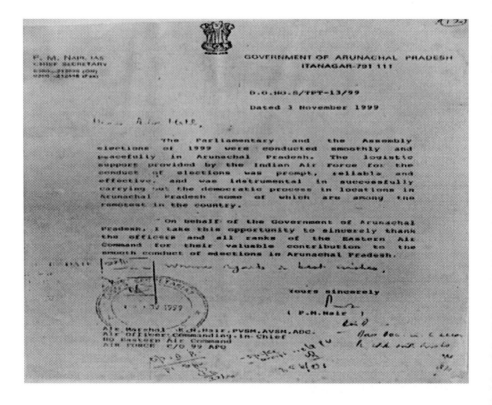

Just Another Day (s)   That helicopter pilots by necessity need to be multi-skilled and adaptable is a truism demonstrated on a daily basis across the country. What better example than the crew, including the CO, of 128 HU on 1st Nov 1999. The helicopter was airborne at sunrise from Tawang for Orissa for the super-cyclone relief. Bad weather was encountered abeam Tezpur, forcing the helicopter to land there and wait. By about 0730 hrs, HQ Eastern Air Command sent a message tasking the crew with a most unusual requirement. A Mig-21 crash a few days back had thrown up a unique challenge. The engine, the prime suspect of the crash, was to be retrieved to investigate for faulty maintenance by OEM and warranty invocation.

However, it was located in the densely forested Kaziranga Sanctuary, half-buried in a completely inaccessible ravine. A plan was made, and the captain along with an engineering officer and few air warriors were winched down by a Chetak close to the site. After a suitable trek, the team reached the site. Some useful out-of-the box thinking and innovative suggestions allowed the engine to be wrapped around with steel cables (from the MT section). A few hours later the Mi-17 approached the site and carried out an OGE hover since it could not come lower than 30 metres due tall trees. Very unconventionally, three sets of 10 metre cables were joined, and the engine virtually prised out of the ground by the powerful Mi-17. It was dropped on to a vehicle about five km ahead at the nearest kutcha track. By the time the crew landed in Tezpur, it was sunset. After a well deserved rest, the next day they reached Bhubaneswar by afternoon. Two days of flood relief later, the same crew was tasked for a retrieval of a Pilotless Target Aircraft (PTA) from the deep sea– a demanding task in itself. So there it was– from high altitude air maintenance to unconventional under slung to flood relief in inclement weather to PTA retrieval. While the CO was an experienced pilot, it must have been a great learning experience for the young flying officer doing his first operational stint. Such are the business-as-usual demands on the fleet!

Nation-Building: IAF Helicopters to the Fore

Airlift for Border Roads 1999-02   China has been our main adversary since the Sino-Indian War of 1962 and Arunachal Pradesh has been critical since then. As a part of strategic plans, a lot of emphasis is laid towards development of roads along the Chinese border. This enables the Indian Army in quick and efficient deployment, should the Red Dragon ever mess with us again. Road building in the Northeastern states assumes high importance due the increasing strategic importance of the area. Often, the necessary equipment for the construction of roads is unable to reach the sites due to vagaries of terrain and weather. The only feasible option left is to get the equipment air lifted by IAF. Mohanbari being a premier helicopter base in the Northeast was once again given the responsibility to provide a much needed helping hand to the Army and the Border Roads (BRO) to achieve this feat, by airlifting bulldozers, trucks, road rollers and all necessary equipment. The Siachen Tigers with their past experience of airlifting guns in the Siachen Glacier were assigned this task in 1999.

A major reason for carrying out air logistics anywhere in the world is because of non-connectivity by road. BRO along with the engineers of the army have the responsibility of building these roads in border areas which helps speedy deployment in war. The Mi-17 is a versatile machine which is capable of carrying load inside as well as underslung. The bulldozers of BRO which can cut through mountains also weigh more than a Mi-17. After detailed planning and referring to all aircraft charts and manuals, a plan was made. The bulldozers were dismantled to smaller parts to get them within limits of the aircrafts capability. Missions were carried to the remotest of helipads, in the most treacherous terrain, some even without helipads. A total of 24 bulldozers, four trucks, one road roller and nine air compressors were airlifted in an effort to link these inaccessible regions by road in the first year itself.

For example, in Nov 2011, BRO had a requirement to airlift a bulldozer to Lee helipad located in the far reaches of Koloriang district of Arunachal Pradesh. The task was initially given to a private operator. The operator, on assessment of the helipad and the narrow valleys restricting the manoeuvring space, wasted no time in pulling out. The task was now assigned to the Siachen Tigers who had a great reputation of having undertaken similar missions in the past. The bulldozer had to be picked up as an external load from Huri helipad and delivered to Lee. The helipad is extremely restricted in space and is famous for posing difficulties even during routine air maintenance missions. This time around, however, the mission was underslung and thus the challenge was greater. Utilising their vast experience, the crew completed the mission with professionalism. Similarly, in Feb 2008, equipment was to be lifted from Kibitoo which is the nearest road head to Gompa which is across the Lohit River and right on the LAC.

Op-Fort William   In Jan 2004, a message was received that the unit would soon be launched for an operation to flush out ULFA militants from Bhutan. The "soon" was actually immediate and aircrew reached the unit before sunrise the next morning for a briefing. The DSS had been up all night

A Tata-407 Truck to a Newly Constructed BRO Helipad

preparing helicopters for the covert mission. One aircraft was launched at first light to Hashimara which was to be the launching base. The CO led another team to Tezpur to setup a second detachment there. Four more aircrafts reached Tezpur the following day. This detachment was being setup for the mother of all missions – Op Fort William. Another detachment was activated and setup at Misamari. The problems faced were many, but the immediate tasks were setting up an Operations room, DSS and establishing aircraft security. Next morning, Kaul and his co-pilot were off to Hashimara. Barely had they landed that the terrible news of the Bhutanese army receiving tremendous casualties reached them. In true Tiger spirit, they immediately got airborne for casevac missions from helipads in Bhutan.

The operation continued despite poor weather and unfamiliar terrain. Lack of intelligence was a major hurdle which the Tigers overcame by meticulous planning. Captured militants were to be transported from Bhutan and handed over to the Indian Army. Aircraft from Hashimara, Tezpur and Misamari also undertook these missions extensively. These militants included the bigwigs of the dreaded ULFA and the KLO. 128 HU flew a total of 30 hours towards Op-Fort William. The effort put in by the unit in mobilising within such a short period and undertaking such demanding missions with professionalism was appreciated by the Bhutanese Army as well as the Indian Army's Eastern Command.

Lt Gen Lam Dorji, DST, DYG, DW

Headquarters
Royal Bhutan Army
Trashichhoe Dzong
Post Box No. 181
G.P.O. Thimphu-Bhutan

DO No. 530/1/A

25 January 2004

Dear Madan,

1.   I, on behalf of all ranks of the Armed Forces of Bhutan and my own, would like to place on record our deep appreciation and gratitude for the prompt and unfailing air support extended to us during the recent operations against the militants. Your officers and men did an excellent job in evacuating the wounded to the hospitals and recovering bodies of those killed in action. They undertook the rescue missions under adverse weather conditions often at risk to their personal safety. These daring officers and men of Eastern Air Command not only saved the lives of the sick and wounded but also contributed immensely in maintaining the morale of the Royal Bhutan Army at an all time high. It also amply demonstrated the enduring close friendship, goodwill, understanding and cooperation that the Royal Bhutan Army enjoys with the Eastern Air Command.

2.   I am taking this opportunity to send one of my staff officers Lt Col Dorji Khandu to personally hand over this letter as a token of our deep appreciation and gratitude for your invaluable support and cooperation.

With warmest regards.

Yours sincerely

Air Marshal MB Madon, PVSM, AVSM, VM**
AOC-in-C,
Eastern Air Command,
Shillong.

Op-Madad   On the fateful night of 26th December 2004, when the entire nation was sleeping an earthquake measuring 9.1 on the Richter scale struck, with its epicentre in the Indian Ocean. It was the sixth deadliest earthquake in recorded history. This resulted in a worldwide catastrophe never seen before in the Asian Sub-continent. A massive Tsunami struck Andaman and Nicobar and the Indian peninsula early in the morning. People were caught unaware. More than 65,000 confirmed deaths were witnessed in the country alone. Air Force Station Carnic was devastated beyond recognition and 102 Air Force personnel and families lost their lives. A giant wave of sympathy swept the country, aptly fuelled by the media, relief measures were planned on a large scale and helicopter units rushed to the South for assistance.

Mohanbari was unaffected by all this, but everyone was rearing to go and reinforce the relief forces. On the 31st, orders were received to mobilise three helicopters from the base for ferry to Carnic via Burma. One aircraft was to go from the unit. The team from Mohanbari was led by Kathpal. New Year celebrations were called off, rightly so, in remembrance of the thousands who perished in the tsunami. The team although new to the terrain, put in their bit and in fact flew more hours and lifted more relief material than any other helicopter in the operation. They flew a total of 69 missions towards Op-Madad in the Andaman and Nicobar Islands and airlifted a total of 71,550 kg and 125 passengers and two casualties. What distinguished the Siachen Tigers from the rest was the silent commitment towards the task and unquestioning sense of purpose which drove them in the hard conditions, weather and fatigue when many were running around making an impression in front of the cameras and entertaining reporters. No rewards were received nor expected but what remained was a sense of pride and satisfaction. The Carnic detachment operated till the 26th of January, a full month after this devastating tragedy struck the nation. The Tigers returned back with their heads held high.

Artillery Guns to Taksing   In 2008, Siachen Tigers were tasked with this challenging mission. Though the Tigers were not new to such missions, this time the airlift was to Taksing, one of the most challenging helipads in the region due to its location. The cumbersome artillery guns were underslung from Mohanbari and successfully dropped at Taksing.

Anti Naxal Ops in Andhra Pradesh   In Jun 2008, 128 HU was tasked to carry out flood relief operations in Orissa and West Bengal. It flew a total of 33 Hrs and supplied 74,000 kg of essential relief supplies to the flood hit districts of both the states. The crew on their way back to Mohanbari stopped at the Air Force Base of Kalaikunda for a night halt. As they sat over a drink, discussing the highs and the lows of the flood relief operations just concluded, they received a call to proceed to Vishakhapatnam (Vizag) the next day. The task was to recover a party of policemen engaged in gun battle with the Naxalites in Malkangiri district. Out of the team of 62 only eight of them survived while continuing to be surrounded by the heavily armed guerrillas.

The crew was new to the area and had very limited safety equipment onboard. The mission was carried out without bullet proof jackets and armour plating. The inherent risk in carrying the mission in the prevailing circumstances was offset by the outstanding planning and flawless execution of the entire operation. The mortal remains and injured survivors were airlifted to safety and the commandoes recovered. At the end of the day, they had flown 16 high risk missions and accomplished a task worthy of a solemn entry in the history books. Little did they know that the sortie would be a first of sorts in the ongoing anti-Naxal operations.

Walong Fire   The Lohit valley, one of the most beautiful stretches of land in the upper reaches of the Indian Northeast, has its own set of surprises to throw at the armed forces deployed there due to the high strategic importance of the area. A major fire broke out close to the army ammunition dump at Walong and threatened to engulf the important army cantonment in no time. The fire started at late afternoon hours and a panic call was received by the Flt Cdr at 1600 hours. The task was to airlift fire extinguishing material and fire fighting personnel to Walong ALG. Sunset was fast approaching and due to the treacherous terrain, unpredictable weather and the approaching night, a swift and meticulous execution was required. The responsibility of carrying out this challenging mission was taken by the Flt Cdr, AP Singh. Because of his vast experience in the area, Mor was selected as co-pilot for this critical mission. The mission had to be carried out on night vision goggles as it was a moon-less night. The aircraft was prepared swiftly by the DSS staff and an exhaustive briefing on ground followed. The crew took off after sunset for Dinjan, quickly took on board the fire fighting crew and equipment, and proceeded to Walong. Flying conditions en-route proved to be worse than expected as the clouding had made the night darker than anticipated. The crew negotiated the route with ease, owing to their sound piloting skills and experience, and landed at Walong ALG safely. By this time the fire was just a few metres away from the ammunition dump. A timely supply of equipment and personnel saved the day for the cantonment in grief and proved the mettle of the Tigers once again.

Recognition by Arunachal Pradesh The state of Arunachal Pradesh (erstwhile NEFA) is a crucial state of the Indian Republic. Its strategic location vis-à-vis China makes it a focus of attention of the country. IAF helicopters and transport aircraft have over the decades contributed in no small measure to the development and well-being of Arunachalis. On 26th Jan 2001, the State Govt acknowledged this by bestowing the Arunachal Pradesh Gold  Medal on the Commanding Officer of 128 HU. The citation acknowledges the selfless devotion to the people of the state when the CO was personally responsible for rescuing and winching up 227 people from the swirling waters of the Siang during the unprecedented flash floods of Jun 2000. The citation also acknowledged the pioneering and herculean efforts made by 128 HU led by the CO in assisting BRO to make roads in the most remote and treacherous terrain of the state.

128 HU was declared the '**Best Combat Unit in the IAF**' on 8th October 2000. The CO and MWO H Sandhu received the trophy on behalf of the unit from the Chief of the Air Staff at the Air Force Day Parade.

PROFILES IN SELFLESS SERVICE:
GALLANTRY AWARD CITATIONS

Flight Lieutenant Harpal Singh Ahluwalia (9821) Flying (Pilot)

On 22nd October 1974 Flight Lieutenant Harpal Singh Ahluwalia (9821) was detailed to undertake special heliborne operations of troop induction against the underground hostilities in Nagaland. He airlifted 60 troops and 2000 Kgs of stores to far flung helipads in two days of intensive flying despite being obstructed by adverse weather. Subsequently, when the unit was again called upon to carry out heliborne operations during December 1974 – January 1975, he airlifted 106 troops and 4180 Kgs of equipment. He flew a total of 195 hrs in 260 sorties which resulted in the capture of a large number of hostiles.

Flight Lieutenant Bagia Laxmi Kanth Reddy (11305) Flying (Pilot)

Flight Lieutenant Bagia Laxmi Kanth Reddy (11305) was commissioned in the Flying (Pilot) Branch of the Indian Air Force in December 1967. He has flown a total of 2251 accident free hours on helicopters over a myriad of terrain. On 22nd October 1974, he was detailed to undertake special heliborne operations, for extensive troop induction in Nagaland against underground hostiles. He undertook this challenging task and within a period of seven days and airlifted 404 fully equipped troops and 9200 Kg of supporting stores to far flung helipads. Regardless of his personal safety he flew 49.05 hours in 122 sorties with undeterred zeal. These operations resulted in the capture of a large number of hostiles.

Flight Lieutenant Bharat Raj Lohtia (11419) Flying (Pilot)

Flight Lieutenant Bharat Raj Lohtia (11419) was commissioned in the Flying (Pilot) Branch of the Indian Air Force in December 1967. He has been serving with a Helicopter Unit since September 1972. During October and December 1974, when the unit was called upon to undertake a massive troop induction in support of combing operations in Nagaland, he was captain of one of the helicopters engaged in the operations. Due to urgency of the situation and limited time, the helicopter had to be flown in the overload variant. The task required courage and flying skill of very high order due to turbulent weather conditions in the inhospitable terrain of Nagaland and ever present threat of small arms fire from the hostile elements. Disregarding his personal safety, he carried out 22 successful missions often flying from dawn to dusk and airlifted a total of 58 troops on 19th October 1974 and another 120 fully equipped troops during the airlift on 31st December 1974. These operations resulted in the capture of a large number of hostiles.

Sqn Ldr Sanjeev Sharma (20489) Flying (Pilot)

On 23rd Jun 2000, an emergency message was received from the Deputy Commissioner, Tezu District that the lives of more than 100 people were at stake as the Lohit River had risen to dangerous levels inundating large areas. Also, a bus carrying 42 passengers was stranded on a remote island. Unless rescued immediately, all 42 passengers were likely to perish. The weather was marginal with incessant drizzle, low clouds and poor visibility. Sqn Ldr Sharma immediately swung into action. Using his skill he negotiated adverse weather to reach the site. Displaying a keen sense of professionalism and skills he winched up five people who were perched on the roof of the bus. Thereafter, displaying his indefatigable spirit, grit determination, he flew 12 sorties rescuing 141 people including six infants from various marooned villages. The rescue operation was so well planned and efficiently executed even in trying circumstances that the casualties were landed at Tezu ALG even before the district administration could reach there. This daring operation enhanced the image of the IAF among the people of the region.

Wing Commander Shashi Kant Mishra (19570) F(P)

Wing Commander Shashi Kant Mishra (19570) Flying (Pilot) is on the posted strength of a Helicopter Unit since 27th Apr 2003 as Flight Commander. The Unit part in 'Op Fort William' in Bhutan, during December 2003. On 15th December 2003, Wing Commander Mishra was tasked to evacuate injured Royal Bhutanese Army jawans in the operations against terrorists. Wing Commander Mishra, gauging correctly the urgency and gravity of the situation, acted in a swift and professional manner. He planned and executed the casualty evacuation himself within fifteen minutes of the receipt of the message. Unmindful of risk involved, he undertook the sortie in hostile conditions and under small arms fire of the militia. His prompt and timely action not only saved the lives of seven critically injured soldiers from Nanglam helipad in Bhutan but also boosted the morale of the soldiers of a friendly country. The professionalism and courage displayed by Wing Commander Mishra has enhanced the image of the Indian Air Force.

VIGNETTES OF SELFLESS SERVICE

While the World Sleeps, Units Prepare for Daily Action; Base Camp OPMD

Tsunami Rescue by Mi-8; Countless Lives Were Saved

Getting Ready for a Night Army Casualty Evacuation

Mumbai 9/11: Nariman House Action by NSG

The Ultimate Test of Skills & Guts: The Siachen Glacier

Special Forces Preparing for Action

CHAPTER IX

Aid to Civilians and Military Duty: A Balancing Act

Introduction   IAF helicopters are one of the most in demand by all agencies, including the Indian Army. Despite good growth over the years, inevitably, demand has always outstripped supply. Even when private players have stepped in a big way, the 'difficult and dirty' or riskier jobs are always left to the IAF helicopters. BRO task at high altitude and helicopter flying in anti-naxal operations of Home Ministry are just cases in point. A great spin-off of all these varying demands has been the on-the-job operational training of young pilots. The sense of independent decision-making, combat orientation, adaptability and mental flexibility are just some of the desirable end-products of such operations. It is not for any other reason that IAF helicopter expertise today is considered one of the best in the world.

This chapter is an attempt to chronicle and document this very facet. The service has to constantly balance these demands and factor in dedicated training that is a must to firm the basic foundations. In the process the nation benefits, since no better optimisation and utilisation of such versatile assets can be hoped for. Snippets and snapshots from various unit diaries gives a window for those outside to understand the complexities of operations in any helicopter unit. Unlike the fixed wing, there are no fixed templates – only uncountable variables. Experience is a key factor in preparing aircrews and this is kept in mind when rotating them from one unit to another. At any given time, the IAF is always prepared to man the huge inventory of heavy-lift, medium-lift, utility and attack helicopters in virtually all scenarios that could come up.

In all this, the basic promise of being prepared for war or to conduct a joint campaign with the Indian Army is never kept out of sight. Each diary extract highlights some of these training endeavours. Anti-naxal operations, currently on in many eastern states, briefly document some of the enabling acts of IAF helicopters. Central Armed Police Forces (CAPFs) rely almost

A Salute to the Indian Nation

completely on this support to make headway in a truly vexing issue – both politically and militarily.

Some Superlative Examples

Rescue at Timber Trail (152 HU)   It was a unique rescue operation undertaken by the unit; and it shot to international fame following the success of this mission which saved 10 lives. On 12th Oct 1992, a cable car of a holiday resort at Parwanoo (Himachal Pradesh) got dislodged and hung mid-air with passengers trapped inside. The unit was tasked to rescue the stranded passengers. The mission was led by Flt Lt P Upadhya. The helicopter winched out all the passengers safely out of the cable car by carrying out a free air hover with absolutely no visual reference for the pilot.

The operation is an example of the live winching capability of the helicopter and the outstanding prowess of flying skill of the unit pilots. They flew a total of 26 sorties clocking six hours to evacuate the tourists.

The cable car had got stuck approximately 1500 feet above ground in the hills of Himachal with ten tourists on board. The ever ready and gallant crew of the unit hatched a risky but doable plan in conjunction with army commandos. The plan was fit to be in the best of Hollywood action movies. A helicopter was hovered with precision over the stranded and precariously dangling cable car. An army commando was winched down on top of the car, who in turn connected each tourist to the winch and was the last one to be pulled up in the helicopter. This daring rescue, the first of its kind in India, was very skilfully executed and covered extensively by the media. The crew were also aptly rewarded with two Shaurya Charkas, one VM and three commendations by the CAS. Apart from these acts of valour that have made 152 HU a jewel in the crown of the air force, the unit has also done innumerable SAR missions, provided aid to civil administration during times of natural calamities and disasters like the floods in Himachal in Sep 1998, 2000 and Jun 2005. The unit was again in the forefront during the Bhuj earthquake relief operations. 'The Armour' routinely carries out exercises like Combat SAR, Special Heliborne Ops, STIE (Small team insertion and extraction), helo-casting (both with Special Forces) along with the regular operations of Air Logistics, communication, VIP flying and Casevac.

Op Black Tornado/ Mumbai 26/11 On 26th Nov 2008, Mumbai was struck by the worst terror attack the country had ever witnessed. Terrorists opened up merciless firing at various places and took over a thousand hostages besides killing hundreds. When the situation spiralled beyond the control of local authorities, the NSG were called in to handle the situation. The Mumbai 26/11 attacks and carnage saw 122 Helicopter Flight based at Air Force Station, Mumbai being very actively involved in assisting and enabling the NSG to flush out terrorists during Op Black Tornado.

The mission was to slither down a team of NSG Commandos on the terrace of Nariman House, where terrorists were holding some Israeli nationals hostage. The crew led by the CO undertook a first-of-its-kind mission to slither down NSG commandos over Nariman House. Unit personnel worked overnight to prepare the helicopter for this mission at a short notice. Mission planning was done after deliberating in detail on various key issues like the location of the windows, balconies, position of friendly troops and direction of fire of terrorists. The helicopter, after getting airborne on time, followed a tactical routing to avoid detection and to maintain the element of surprise. The roof or the drop zone was a small area of approximately 15 feet × 15 feet surrounded by very high density of obstructions such as tall pipes, wires and a concrete wall. This being the case, the situation precluded the possibility of a low hover jump or a low height slithering. The slithering had to be planned at a minimum height of fifteen metres and thereafter aim to reduce the height after assessing the situation on the scene. This was the only way the troops could have been transferred, even though it exposed the helicopter to hostile fire during the hover. When it arrived on scene, the situation was worse than anticipated with restricted hovering space and direct exposure to enemy fire. The slithering of 21 troops in the first wave was carried out in the restricted and confined area, amidst the sound of incessant firing and exploding grenades.

The helicopter carried out two more slithering sorties to complete the full complement of NSG troops for further operations, even though the element of surprise was lost for the subsequent two sorties The mission undertaken to slither NSG Commandoes on the roof of Nariman House was extensively covered by the media in India and abroad. It had once again proved IAF's mettle, professionalism and high heroic traditions. In view of the safe and excellent handling of the mission, the crew were conferred Vayu Sena Medal (Gallantry) by the President of India.

Anti-naxal Operations: Supporting CAPFs

In Jun 2008, heavy rains flooded the plains of West Bengal and Orissa. The civil administration of both the states asked for help. 128 HU was tasked and the next day Ray along with Kulkarni was detailed to carry out flood relief operations in Orissa and West Bengal. They flew a total of 33 hours and supplied 74,000 kg of essential relief supplies to the flood hit districts of both the states. The crew on their way back to Mohanbari stopped at the Air Force Base at Kalaikunda for a night halt. As they sat over a drink, discussing the highs and the lows of the flood relief operations just concluded, they received a call to proceed to Vishakhapatnam (Vizag) the next day. The task was to recover a party of policemen engaged in gun battle with the Naxalites in Malkangiri district. The heart wrenching news was that only eight of the party of 62 policemen were alive and that too with severe injuries. Naxalites had become active in the state since the past few months. These affected areas in the state had a high concentration of police force deployed to counter the menace. On that fateful day, 62 policemen were travelling in a boat near Balidela dam in Malkangiri district to augment the fighting force at a police post. The party was ambushed and came under heavy fire. The boat they were travelling in had suffered heavy damage and capsized. The firing continued, and most Jawans perished to bullets or drowned. Out of the team of 62 only eight survived while continuing to be surrounded by the heavily armed guerrillas.

On reaching Vizag, Ray was informed that the Navy had refused to launch their Sea King helicopters citing lack of experience in this kind of mission and the fact that the area of operation was not sanitised. The mission was urgent and required swift decision making. Ray, a Helicopter Combat Leader, assessed the threat with the limited intelligence inputs available. He chalked out a strategy in which the plan was to insert CRPF 'Grey Hound' Forces and Naval Marine Commandos, who would sanitise the area and provide a safe landing zone for the helicopter to land and recover the survivors and the bodies of the deceased policemen. The crew was new to the area and had very limited safety equipment onboard. The mission was carried out without bullet proof jackets and armour plating. The inherent risk in carrying the mission in the prevailing circumstances was offset by the outstanding planning and flawless execution of the entire operation. The mortal remains and injured survivors were airlifted to safety and the commandos recovered. At the end of the day, the Tigers had flown 16 high risk missions and accomplished a task worthy of a solemn entry in the history books. Little did they know that the sortie would be a first of sorts in the currently on going 'Op-Triveni'.

On 14th Nov 2008, a crew of 119 HU was tasked to carry out election duties at Jagdalpur in Chhattisgarh. It flew 58 sorties airlifting 215 passengers and 2905 kg load in the Naxalite affected areas. In keeping the democratic process of elections alive, unit lost one air warrior when the helicopter came under fire from the Naxalites during takeoff. The crew showed exemplary courage and presence of mind taking evasive action and despite the loss of one crew member and a damaged aircraft, flew back to safety saving the lives of 12 passengers. This act also denied a source of propaganda to the anti-national elements and projected the image of the unit and the Air Force in a praiseworthy manner. 152 HU operated a four helicopter detachment at Jharkhand for election duties in Nov and Dec 2009. 475 sorties were utilized for airlifting 16 tonnes of load and 3741 passengers. Apart from the main role of airlifting polling parties, it carried out numerous casualty evacuations and area domination sorties in Naxalite affected areas.

As the threat from Naxalites continued to spiral upwards in the country and turned out to be the single largest internal threat to the country by the end of 2009, IAF was tasked to support counter-insurgency operations in coordination with paramilitary forces and civil police in the Red Corridor. As the unit had continuously proved its mettle in special operations, it was tailor made to suit the requirements demanded in this hostile environment. Four helicopters of this unit were deployed at Raipur and Jagdalpur from Dec 2009 to Sep 2011. Operations were undertaken in a large geographical area comprising of eight states, thereby making it one of the largest helicopter operations ever undertaken by IAF. All helicopters were configured with door and blister mounted guns manned by IAF commandos or Garuds. The unit also undertook the training of paramilitary forces so as to carry out seamless coordinated missions. It flew 2566 sorties clocking 1746 hours; inducted 17894 troops, evacuated 88 casualties, brought home 163 dead bodies and airlifted 299 tonnes of load. The unit also trained 984 paramilitary troops in slithering operations. The operation brought the intangible skills like jointmanship and interoperability to the forefront and left an example for other units to follow. The mere presence of the helicopters has drastically cut down the attrition rate of the paramilitary troops and boosted the spirits of the ground forces in the insurgency hit regions.

During the last five years, Naxalism has grown to be a substantial menace, threatening the internal security of our country. Naxalites take advantage of the disenchantment prevalent among the exploited segments of the population and seek to offer an alternative system of governance through the barrel of a gun. As the problem became unmanageable, the Government of India launched a counter-offensive known as 'Operation Green Hunt', which called for the deployment of more than 50,000 paramilitary soldiers to the regions that are most affected by the increased violence of the insurgency. The hostile terrain and the alien conditions led to increased casualties of paramilitary troops during routine movements. It was then that the Government realized the overwhelming need of IAF's expertise in such roles. 152 HU has now been replaced with 105 HU in these operations.

Panning across eight states, Op Triveni is active in Chhattisgarh, Bihar, Jharkhand, Orissa, West Bengal, Madhya Pradesh, Maharashtra and Andhra Pradesh. It is one of a kind helicopter operations conducted in a sustained manner by Indian Air Force. The unit has been a pioneer in establishing support operations for civil police and paramilitary forces in the never-treaded Red Corridor. All this was accomplished within a strict realm of Rules of Engagement, restricting any offensive air action. The operation was intended to carry out peace time support missions like air logistics, casualty evacuation and troop insertion/ extraction. The tactics, techniques and procedures to operate in such a scenario were worked out from the scratch and have proven to be quite effective till date. In a bid to bring in jointmanship and inter-operability, it had trained 984 personnel of police and paramilitary troops for slithering operations. It had also conducted ground training programmes for paramilitary forces and ensured seamless integration of Garud troops in real-time operations.

The principal accomplishment of the IAF in this operation has been in achieving good understanding with civil and paramilitary forces. It had identified that conspicuous team effort was a key for successful counter insurgency operations and hence dedicated a large part of its effort towards this cause. The unit has always been prompt in assisting the civil police and paramilitary forces for which it had received several letters of appreciation from these agencies. The employment of helicopters for anti-Naxal operations had been a huge morale booster for ground troops employed on this role. There had been a drastic reduction in the attrition rate of paramilitary forces as a result of deployment of IAF helicopters in Operation Triveni. In spite of the hostile conditions in the mission area, it accomplished every mission with utmost professionalism.

As the distance between the parent base and the detachment location was very large, it was decided to carry out second line servicing of the helicopters at Raipur. This was one of most remarkable achievements of the unit as it was for the first time that such an activity was being performed at a non-IAF base. The contribution towards anti-Naxalite operations has been in a sustained manner bringing laurels to the Indian Air Force. All this has been achieved while operating under the most hostile and hazardous of environments. The Mighty Armour, in the operations has lived up to the motto of 'Who Dares Wins'. Some snapshots of operations follow.

Jharkhand Elections (22nd Nov to 20th Dec 2009)   The Unit ensured free and fair elections in Jharkhand and Bihar by positioning the security personnel in the sensitive constituencies and carrying out regular air surveillance mission. As a result, for the first time in the recent times, 60 percent of the people exercised their voting rights.

Dantewada Massacre (6th to 8th Apr 2010)   In the wee hours of 6th Apr, Maoists attacked No.2 Battalion of CRPF camp based at Chintalnar in Dantewada District of Chhattisgarh state. More than 75 troops were brutally killed by the Naxalites. Two helicopters got airborne from Raipur within the shortest possible time and evacuated the martyrs to Jagdalpur. They continued to induct more troops near Chintalnar which obviated any further attacks. The unit flew 26 sorties, clocking 21 hours and airlifted 184 passengers, 64 martyrs and 2.3 tonnes of load towards this incident. This horrific incident sowed the seeds for a new detachment at Jagdalpur.

Train Accident (28th to 30th May 2010)   At around 0120 hrs on 28th May 2010, Maoists blew up the rail track near Kalaikunda, West Bengal. This resulted in derailing of Gyaneshwari Express. The freight train which was following rammed into the derailed train resulting in death of more than 150 passengers. One helicopter evacuated seriously injured passengers within the golden hour, saving many precious lives. The unit undertook six hours of flying and airlifted ten seriously injured patients. The timely assistance provided was deeply appreciated at all levels.

Dhaurai Massacre   Maoists attacked the CRPF post at Dhaurai on the late evening hours of 29th Jun 10 during which 27 troops were killed. Two helicopters were launched for Casevac from Raipur and Jagdalpur early in the morning of 30th Jun 2010, while the third aircraft was carrying out airborne patrol. After the evacuation of all the injured troops and martyrs, fresh troops were inserted employing two helicopters. The unit undertook five sorties clocking five hours and airlifted 74 passengers and 26 martyrs.

Cas Evacuation Ex-Bheji helipad   A Casevac mission was carried out on 13th Sep 2010 from Bheji helipad where fierce gun battle was going on between CRPF and Naxalites. It was here that the helicopter was fired upon by Anti-National Elements (ANE's) while taking off from Bheji. The crew through sheer grit and presence of mind got away and saved a possible embarrassment at the hands of Maoists. 20 injured CRPF personnel requiring immediate evacuation from the hostile and inaccessible area were successfully flown to the nearest medical centre.

Grey Hound Ops   A massive ground operation was launched by the Special Forces of Andhra Pradesh police – 'The Greyhounds' on 17th Dec 2010. This resulted in killing of four Maoists and capture of another five. However, the Grey Hounds were stuck for recovery for want of ration and ammunition, thus endangering their withdrawal. Two helicopters were launched from Jagdalpur for extraction of more than 130 Grey Hound troops from thick jungles of Andhra Pradesh north of Narayanpattnam. The operation was carried from make – shift clearings in the jungles. The mission was a huge success in boosting the morale of the greyhounds.

Pre-emptive Ops (18th Dec to 19th Dec 2010)   In one of the most proactive moves against Maoists, the state government of Chhattisgarh planned a pre-emptive attack against the Naxalities in Jagargonda and Chintalnar sector. This was based on concrete intelligence reports of an imminent attack on these posts. IAF helicopters were quick to respond and inserted 180 troops of CRPF Cobra Bn and two tones of ammunitions and ration, thus thwarting a Naxalite attack. The mission turned out to be a great success and the unit was lauded by all the CRPF dignitaries.

Security and Sanitisation Ops (1st to 3rd Apr 2011)   In a major bid to achieve security and sanitization operations by the CRPF and Para-military forces prior to Chief Minister's visit in Chintalnar area, IAF flew 36 hours, inserting 578 security men and six tonnes of ammunition and rations at Chintalnar and Jagargunda axis.

Recovering the CAPF Martyrs; The Quick Counter-Punch

Squadron Leader TARUN KUMAR CHAUDHRI
(25871) Flying (Pilot)
Shaurya Chakra

On 14 Nov 08, Squadron Leader Tarun Kumar Chaudhri was assigned the task of picking up three Border Security Force casualties, Electronic Voting Machines (EVMs) and election commission officials. He was chosen as the captain of the helicopter because of his leadership, courage and qualities of task accomplishment. He proved his mettle and worth by successfully undertaking the task at Chatishgarh, immensely contributing towards the planned election process in the state in spite of operations involving regularly flying in the disturbed areas of Chatishgarh due to violence of Anti-National Elements (ANEs). On 14 Nov 08, after landing at the helipad and picking up the personnel and equipment, during take off, his helicopter came under heavy small arms and light machine gun fire. The helicopter sustained heavy damage to the main rotor blades, fuel tank, and tail boom. Despite damage to aircraft and fast approaching night fall, Sqn Ldr Chaudhri displayed supreme courage under fire, skilful handling of the aircraft and grate composure in quickly and correctly assessing the fly-worthiness of the aircraft, the adverse ground situation due to the presence of ANEs in the vicinity and flew back the helicopter safely to Jagdalpur.

Squadron Leader Tarun Kumar Chaudhri displayed exceptional courage under fire from anti national elements, leadership and composure under adverse situation. For this courageous act he was awarded Shaurya Chakra.

K. VIJAY KUMAR, I.P.S.
Director General
CENTRAL RESERVE POLICE FORCE
Block I, C.G.O. Complex
Lodhi Road, New Delhi - 110 003
P : 24360971 Fax : 24363192

D.O No. T.VI-1/2012-Ops-2   Dated, the 23 March, 2012

Dear Air Chief Marshal N A K Browne .

    I can't tell you how much we, at the Directorate General, CRPF, value the stellar contribution made by the IAF in a very iconic ops that was recently launched in Chhattisgarh. This ops was very unique in that an unknown turf was being opened up, by over 2500 men on foot over a 02 weeks period.

2.   However, there was an effort on 16/03/2012 to derail the heli ops near a place Jatwar - right in the heartland of this territory unexplored by SFs, which was very ably countered by Flt. Lt. Nikhil Mehrotra and Flt. Lt. Gaurav Lohani and other crew. Their brilliant intervention is greatly appreciated.

3.   We would like to once again express our sense of gratitude for the Indian Air Force which under your able command has been so forthcoming and supportive in all our CI efforts. We look forward to more of the same in the days ahead.

    With regards,

    Yours Sincerely,

    (K. Vijay Kumar)

Air Chief Marshal N.A.K. Browne, PVSM AVSM VM ADC
Chief of the Air Staff,
Indian Air Force,
New Delhi

R.S.H.S. SAHOTA

D.O. No. G.II.1/10-OPS-PA
Dy. Inspector General of Police
Central Reserve Police Force
Dantewada (C.G)
Dated the, Nov' 2010

Dear

You were instrumental in flying DIG (Ops) CRPF, Dantewada, DIG cum SSP Dantewada, Commandant 111 Bn along with 204 Cobra reinforcement team to Jagarguanda against grave risks after fierce encounter wherein we killed 9 Naxals and recovered their bodies.

2.   Further done yeoman service in evacuating two martyrs from Awapalli to Bijapur. I place on record my sincere appreciation for this gallant and noble role.

With

Yours

(R.S.H.S. Sahota)

OFFICE OF THE DIRECTOR GENERAL

CENTRAL RESERVE POLICE FORCE

LETTER OF APPRECIATION

Since December 2009, the presence and employment of helicopters toward Operation Green Hunt has been a real shot in the arm for the Central Reserve Police Force. The quick reaction showed by the pilots in inserting CRPF troops into Chintalnar sector on 06 April 2010, after a horrific attack by the naxalites, acted as a deterrent and evidently obviated any further attacks. The No.2 CRPF Battalion will never forget the moment when the helicopters evacuated the injured and brought home their martyrs. The immediate response of the pilots following the attacks on CRPF post in Dharai region on 30 June 2010 was exemplary and proved vital in boosting the morale of our troops. It would indeed be redundant to mention about several other difficult missions taken up as routine tasks by 152 Helicopter Unit.

The helicopters have acted in a singularly outstanding manner to every challenge placed before us. Your mere presence has been vital in boosting the morale of our troops. The efficient manner in which the helicopters were employed exhibits the truly professional attitude of the officers and men posted at 152 Helicopter Unit. Their efforts during the last one year toward Operation Green Hunt were one of the major contributing factors in achieving many successful exercises against the extremists. The commitment displayed by these men reflects great credit upon themselves, 152 Helicopter Unit and the Indian Air Force.

It is my desire that each individual of 152 Helicopter Unit be commended for a truly professional display in an outstanding manner. I would like to place on record, my sincere appreciation, to 152 Helicopter Unit and best wishes in all future endeavours.

(COPY)

Lt Gen VN Sharma, AVSM
GOC

13048/Ex MS&PA/G(T)

HEADQUARTERS 1 CORPS
C/O 56 APO

24 May 85

1.   I am writing to express my deep gratitude for all the assistance you and your staff so spontaneously gave us to make exercise MAY FEVER and PIERCING ARROW very successful. Without the backup of the resources made available by your Command we would have not been able to introduce the necessary realism as far as the air aspects were concerned. Your pilots and ground staff gave us a conspicuous display of professional competence and efficiency, and we enjoyed having them with us.

2.   I shall be grateful if you could convey my appreciation to all the elements that participated.

Sd/-

Air Marshal DA Lafontaine, AVSM, VM
AOC-in-C
Western Air Command
New Delhi

Squadron Leader RAJINDER PAL SINGH DHILLON
(7741) Flying (Pilot)
Shaurya Chakra

Squadron Leader Rajinder Pal Singh Dhillon (7741) was commissioned in the Flying (Pilot) Branch of the Indian Air Force in November, 1962. He served as Flight Commander of a Helicopter Unit since March, 1974. On 26th January, 1975, he was detailed to carry out the earth quake relief operations in Himachal Pradesh. The weather on this day was extremely inclement and the hostile terrain made flying efforts extremely hazardous. With a deep sense of devotion, relentless zeal and determination, he executed his task with true professionalism. Throughout that day he braved and battled against weather and terrain and delivered his vital loads of relief material and evacuated those in need of urgent medical attention. During subsequent missions, he executed daring landings at unaided helipads and precision drops at miniature Dropping Zones. During these operations he flew a total of 52 stories and airlifted large quantities of material which resulted in timely relief to the affected people.

Squadron Leader Rajinder Pal Singh Dhillon displayed courage, professional skill and devotion in duty of a high order. For this courageous Act he was awarded Shaurya Chakra.

Squadron Leader GULZARINDER SINGH BRAICH
(8762) Flying (Pilot)
Shaurya Chakra

On March 1979, Lahul and Spiti valley in Himachal Pradesh were subjected to heavy snowfall and avalanches which had disrupted all traffic to and from the valley and had considerable loss to life and property of the inhabitants. Squadron Leader Gulzarinder Singh Braich was detailed to undertake relief supply missions in the area. Undeterred by the adverse weather conditions and the consequent flying hazards, he flew eight sorties with 6,700 kilograms of supplies for the affected people.

Squadron Leader Gulzarinder Singh Braich thus displayed exceptional professional skill, devotion to duty, courage and bravery of high order. For his dedication and devotion to duty he was awarded Shaurya Chakra.

J.N. Tripathi,
District Magistrate,
East Champaran.

MOTIHARI
Dated 31 August 1986

Dear Sir,

We are extremely grateful to you for so promptly making available your assistance to us in a moment of crisis when this district of East Champaran has been reeling under unprecedented floods. Around four lac people have been severely affected and about three hundred villages have been marooned by floods. Road and rail communications having collapsed due to onslaught of folds, your aircrafts have been the only and most effective means to provide relief to affected people.

During last four days your officers - Wg Cdr AG Bendre, Sqn Ldr Ajay Sharma, Flg Offr Joe Emmanuel, Flt Lt TPS Dhillon, Sgt Chaudhury Sgt Das, WO Sirkal and JWO Sethi and your other men have worked day night and done immense help to the people. They are leaving behind a clear impression about the efficiency and capability of the Indian Air Force in reaching the people in hour of need. On behalf of district administration as well as the people of this district, I express my heart felt gratitude to you and believe that the same assistance and prompt help shall be available in future again whenever we need it.

Thanking you once again.

(Copy)

S.G. AGRAWAL, IAS

d.o. No. 3111-12/WHC
dt. 20/30 Jul 88

It gives me great pleasure to put on record my appreciation and thanks for the help rendered by your aircrew during the past few days for conducting flood relief op. The effects of your officers and men were of invaluable help to us, inspite of all that was required, unmindful of the discomfort and long hours.

2.   I would like to make a special mention of Sqn Ldr M Dutt and Flt Lt JS Kandar who not only carried out their duties very efficiently but also left a lasting impression on us with their pleasant and courteous behaviour.

Yours sincerely,
Sd/- xxx
(S.G. Agrawal)

Group Captain I.S Bindra,
Station Commander;
Air Force Station;
AdampurC.

Copy to :
1. Air officer Commanding
No. Wing, Air Force,
C/O 56 APO

Balancing the Demands: Unit Legacies

104 HU   The history of 104 HU, the most prestigious and the oldest helicopter unit of the IAF, is studded with more than six decades of yeomen service rendered to the country. It was raised as a helicopter flight at Palam on 10th Mar 1954. The first helicopter S-55 arrived at Bombay by sea on 19th Mar 1954 and subsequently test flown on 23rd March and ferried to Delhi on the 25th. Such swift action, complete involvement and dedication on the part of the crew and technicians was the first clear indication of the standards that 104 HU set for itself. Within 72 hours of the arrival of its first helicopter, the flight undertook its first VVIP commitment on 28th Mar, when it carried the first Prime Minister of India, the Late Pandit Jawaharlal Nehru, from Palam to Tilpat range and back to witness the 'Fire Power Demo'. During the same year, it did its first mercy mission when Flt Lt AN Todd, evacuated 15 villagers marooned in the middle of Yamuna river (near Delhi). A very responsive press started reflecting to helicopters as the 'Harbingers of life'. In May 1955, it carried out flood relief operations in Assam. In September, it heeded to a call of distress and moved a detachment of two helicopters to Barrackpore and Bhubaneswar for flood relief in Orissa. On 1st Jul 1958, two survivors of an IAC plane which crashed in NEFA were successfully rescued. In Aug 1959, a detachment of two S-55 and two Bell helicopters, under the command of Flt Lt SK Majumdar moved to Daprijo (an advanced landing ground in NEFA). The detachment did a marvellous job by evacuating more than 200 Tibetan refugees from Limiking, a small village on the Indo-Tibetan border.

In 1977, it joined the ranks of the fighting forces of the IAF, when the role of the unit was re-designated to an ATGM (Anti Tank Guided Missile) unit. All helicopters were modified to carry four AS-11B guided missiles. It took part in numerous army/air force joint exercise involving tactics for the new ATGM. In Dec 1978, in the Bhagirathi River Valley relief operation, food supplies were airlifted from Sarsawa to Uttarkashi, Harsil and Nelang, and stranded pilgrims were brought back in return flights. In Mar 1979, the unit carried out mercy missions dropping food and medicines in the worst ever avalanche in the history of Himachal Pradesh.

110 HU   The unit has been carrying out major operations with Indian Army in numerous exercises and has become an integral part of army war games in the east. In Sep 2007 unit took part in Exercise Vajra Prahar and carried out slithering of troops and casevac flying 21 sorties. In Oct 2007, it participated in Exercise Shatrujeet, slithering 87 troops of 14 J&K LI in Limakong. During Poorv Abhyas, the unit helicopters operated from Mohanbari, slithering 258 army troops. It has been flying numerous slithering and other combat sorties for the army and has become an integral part of their day to day training. In 2008, it carried out slithering of 690 troops in CIJW School, Wairangte; another 1314 troops in 2009, 866 troops in 2010 and 143 troops in 2011. Apart from this, it trained 87 troops in Haflong and 212 troops in Hashimara in 2011.

Air maintenance sorties are being carrying out for the army on a regular basis; which increased with the conversion to Mi-17 in 2006. It took part in Garud–III (Indo-French exercise and Indradhanush (Indo-UK exercise) held at Kalaikunda by giving ASSR support and carrying out slithering of Indian and French troops. A day after the powerful earthquake hit Sikkim in Sep 2011, the unit mobilised its two helicopters on a war footing to Gangtok. In spite of inhospitable conditions and bad weather, Vanguards were in the middle of it carrying out relief and the National Disaster Relief Force to Lachen, Lachung, Mangan and other interior regions of Sikkim. It carried 150 tons of rescue and relief material to various Army posts flying for 143 sorties/ 102:30 hrs. The Unit continues to carry out PTA recovery and radar calibration sorties over sea. It also maintains detachments at Limakong, Kamzawl and Agartala for air logistic task to Army/ BSF and one helicopter to provide search & rescue cover over sea.

107 HU   In May 1972, the newly commissioned Sagar Samrat at Bombay High had one of its vessels with 22 ONGC personnel missing in high seas. The Unit conducted a successful search and dropped medicines, water and food for the marooned. In the same year, it took part in flood relief operations in Rajasthan. The beginning of 1990 saw it proceeding to Jamnagar to operate for ONGC; flying 187 sorties and 141:45 hrs with two helicopters. The support to Indian Army continued and 28 hours were flown in support of ground forces. In Jul 1990, Desert Hawks were once called upon to undertake flood relief ops in Samdhari, where it flew 40 hours and saved numerous lives in the affected areas. During the second half of the year, support to Indian Army took precedence over all other commitments with 61 sorties in 64 hours. In Jan 1991, the Unit participated in Exercise Shiv Shakti with 50 (Indep) Para Bde. The contingent was led by the CO, Wg Cdr AK Bhardwaj, and the detachment flew 34 sorties and 42 hours. Both day and night slithering of troops was carried out during the exercise. Tasking for

Ex Dayasagar and Agniban towards the end of the year included an effort of 95 sorties and 85 hours. It flew extensively for various agencies like the Indian Army (104 sorties/ 106 hours), Govt of India (411 sorties/ 384 hours), VVIP/ VIP communication duties (86 sorties/ 90 hours) and ONGC (184 sorties/ 142 hours).

During 1992, it was heavily committed with tasking of various ministries and departments of the Govt of India. The tasks included oil-rig ops at Bombay High, ops in Dehradun–Shimla area and flood relief in Gujarat and Rajasthan. This took 224 sorties and 416 hours during the year. In 1993, it was involved in disaster relief ops in various parts of the country which included earthquakes as well as floods. During VVIP/ VIP communication duties, helicopters were utilised as far as Shillong by flying 170 hours in this role. Despite heavy tasking in these duties, the responsibility towards Indian Army was never neglected and close to 160 hours was devoted towards various exercises with the Army. In Mar 2006, it participated in Baaz Shakti for which 45 hours were flown. Another life was saved when an army casualty was evacuated from a post around Jaisalmer. In August, the Hawks proceeded to Vadodara for flood relief, where they dropped 100 tonnes of relief load and carried out live winching of 200 odd persons in just five days. Towards the end of the year, it participated in Ex-Shatruvinash and Ex-Gagan Shakti with the Indian Army.

In Feb 2008, Wg Cdr V Raj carried out yet another night casualty evacuation sortie for Army from Uttarlai. In Mar 2008, it participated in an International Exercise – Brazen Chariot, involving troops from the U.S. 64 sorties were flown towards this exercise which was conducted jointly with the Indian Army. Other than this, it also participated in exercises Prashikshan, Dakshinshakti and Dhramvijay with the army in Agra. In Aug 2008, it again moved for flood relief to Purnea flying 34 sorties and 48 hours. The last exercise for the year was Ex-Rannbankure at Nachna with 35

sorties/ 22:50 hours with the Indian Army. 2009 began with an air logistic detachment at Srinagar in support of army operations flying over 100 hours in 230 sorties in Jan and Feb 2009. The year saw Wg Cdr RN Jha carry out a SAR mission of ejected pilot on NVGs from close to Pokhran. Soon to follow was one more feather in the cap for the Unit when the Commanding Officer led the Unit in trials of GSh guns on Mi-17s in Jan 2010. It carried out extensive flying towards Ex-Mahavayu Shakti with the army totalling 102 sorties and 90:25 hours in Jan – Feb 2010.

2011 began by proceeding to Srinagar for detachment. More than 80 hours were done towards army task. It was thereafter tasked to provide the requisite air effort for the 11 HCL course conducted at TACDE. Exercise Bulls Eye was the next in the calendar, where two helicopters from the Unit flew close to 45 hours with 50 (Indep) Para Bde and carried out various missions like slithering, SHBO, under-slung of vehicles and guns, etc. The last quarter of 2011 saw increased participation in joint exercises with the Indian Army and the conduct of Ex-Mahagujraj and Sudarshan Shakti. It flew 55 hours in 68 sorties training about 1500 troops while operating with the army.

119HU   The Unit has been called upon time and again to undertake missions in aid of civil power. It carried out earthquake relief at Latur in Maharastra in 1993. It was involved time and again in flood relief operations in Gujarat, MP and Rajasthan sector. During the cyclone that hit Gujarat area in Jun 1996, it carried out innumerable SAR missions saving precious lives. It rescued civilians and navy personnel and earned two Vayu Sena Medals. In the same year, it carried out extensive flood relief operations in Gujarat and Rajasthan providing relief and succour to the marooned people. In 1998 flood relief operations in Surat, it rescued 19 people and air dropped 75 tons of relief material to the marooned people. During the 1999 super cyclone, the unit was again put to service. On 19th May 1999, a SOS message was received to rescue personnel from a country vessel off Porbandar coast. One helicopter was launched from Baroda to search for and rescue these personnel. The helicopter successfully winched up eight persons from the open sea under extremely strong wind conditions (60-70 Km/h) sea state 5 and cloud base 300 m. Another helicopter winched up five more persons.

In Sep 1999, it under took flood relief operations in Hoshangabad, and in 2004, carried out flood relief ops in Bihar, Gujarat, Daman and Diu saving precious lives and providing much needed supplies to the civilian population. In May 2004, it rescued six Marine and Para commandos who were lost in the creek area and were missing for 24 hours. It also maintained a one aircraft detachment in Sri Lanka and provided medical relief and casevac to the people affected by Tsunami. In 2006, it launched a successful search & rescue mission for an ejected MIG 29 pilot. The mission was launched at night during dark phase and hence required a very high skill level from the crew. In 2007, it was called upon to undertake a rescue mission in the Union Territory of Diu. The crew displayed exemplary skill in rescuing eight fishermen stranded on a lone tower in the middle of the raging sea in extreme weather conditions. In another instance of rescue, the unit picked up 18 crew members of a ship MVS Glory belonging to Iraq, which had run aground due to bad weather in Kori creek.

The unit is constantly involved with the Indian Army in major and small tactical exercises. It is a frontline unit of IAF's South Western Air Command for all quarterly and annual exercises besides regular slithering training of Indian army units.

In Nov 2008, it was tasked to carry out election duties from Jagdalpur in Chhattisgarh. It flew 58 sorties airlifting 215 passengers and 2905 kg load in the Naxalite affected areas. In assisting the democratic process of elections, it lost one air warrior when the helicopter came under fire from the Naxalites during takeoff. The crew showed exemplary courage and presence of mind to take evasive action despite the loss of one crew member and a damaged aircraft, flying back and saving the lives of 12 passengers. This act of bravery also denied a source of propaganda to the anti-national elements and projected the image of the unit and the IAF in a praiseworthy manner. On 27th Nov 2008, the entire nation was shocked by the audacious terror attack in Mumbai. On receiving a call, the unit immediately launched two helicopters to participate in Op Black Thunder. It successfully dropped 64 NSG commandos for the mission and airlifted 11 passengers and five casualties during the operation.

109 HU 1963-65:   During this period, the unit was employed in multifarious roles in its effort for consolidation, which included among other things, the tasks of flying for Border Roads Development for construction of mountain roads in the difficult terrain of the Northeast. In Aug 1963, Jagadhri, a developing town in Western Uttar Pradesh, was ravaged by floods. The unit with its whirly birds in their new found role as Harbingers of Life, descended from the skies to bring relief to the distressed citizens. In Jan 1964, when riots broke out in Srinagar, reacting with lightning speed, it flew hordes of security men into sensitive areas in the valley and patrolled the skies keeping a hawk like vigil for signs of imminent trouble. Even as all this was happening, it regularly took part in Republic Day fly pasts in Delhi besides being ever ready to rush to the rescue of their comrades at arms (Indian Army) in the Sugar Sector. In the month of March 1965, it commenced operations from ALGs in Gujarat (the Rann of Kutch area). In Jul 1965, massive air drop of food and supplied was carried out for marooned troops in the same area. On 15th Aug 1965, a detachment of six helicopters was positioned at Jammu for operations, wherein they flew with distinction in the Pathankot and Chhamb-Jaurian sector. On 21st Aug 1965, this detachment was moved from Jammu to Srinagar to undertake varied roles like bombing, strafing, casualty evacuation, supply dropping, reconnaissance and communication.

After the 1965 War, unwilling to throttle back and relax, it went into action again this time in the eastern sector. Four helicopters were flown to Mizoram in 1966 to take part in the Mizo Hills operations. The domain of One Zero Nine was phenomenal, as simultaneously a detachment was being maintained at Jodhpur for operational commitments. On 10th Apr 1967, the unit moved from Chandigarh to Jammu leaving a detachment for operations in the Sugar Sector. It proved beyond doubt that helicopters were effective force multipliers both in war and peace, and thereby, highlighted the ever increasing importance of their roles. The inevitable happened in the first week of Dec 1965. One Zero Nine not only rose to the occasion but surpassed all expectations and imagination. The helicopters of 109 HU were the first to land at Jaurian and lift out the first batch of casualties. It flew 315 hours and lifted out a total of 468 casualties directly from the battle zone under enemy fire. In 1986, incidents of Chinese intrusion took place in the north eastern parts of the country, and 109 HU pioneered Op Falcon under difficult conditions of terrain and weather; streamlining the operation and giving the impetus to sustain the Indian Army. The mementos and letters presented to this unit stand silent yet eloquent testimonial of the gratitude and confidence reposed by the Indian Army.

On 26th Dec 2004, an earthquake measuring nine on the Richter scale with its epicentre in Sumatra resulted in a giant tsunami. This caused extensive damage to human life in Indonesia, Thailand, Sri Lanka and India. 109 HU was called on to provide relief in the friendly neighbourhood country of Sri Lanka. The tsunami had devastated the eastern coast of Sri Lanka. Heavy damage to property and great loss of lives was reported from every costal village. On request of the Sri Lankan Government, a team of six Mi-8s led by UK Sharma reached Katunayake-an air force base. IAF helicopters undertook task such as: search for marooned people and survivors; rescue people in affected areas; airlift essential supplies and medical aid. These operations which began with the emergency phase, continued beyond the relief phase and ended in the support and assistance phase. Operating in an international environment alongside soldiers from countries like the US, UK and Bangladesh, it clocked 430 sorties in 270 hrs of incident free flying. More than 235 tonnes of supplies was airlifted and 67 passengers saved during these operations. For the exceptional devotion and dedication during tsunami relief operations, Wg Cdr UK Sharma was awarded Vayu Sena Medal (Gallantry), Wg Cdr V Navad, SEO & MWO Roop Singh Flt Eng were commended by the CAS and JWO TR Saran Inst Fit & non-combatant CB Obalesu were commended by the AOC-in-C, SAC.

111 HU   The monsoons of 1968 were truly monstrous in West Bengal. Jalpaiguri district was submerged and thousands were rendered homeless. Gp Capt LM Katre, Stn Cdr Hashimara and Sqn Ldr BD Dangwal, CO 111 HU AF came to the rescue and launched a massive flood relief operation wherein 57,705 kg and 765 passengers were airlifted. Triple One once again proved to be the harbingers of mercy. On 24th Feb 1969, Sqn Ldr KC Cariappa took over the command of 111HU from Sqn Ldr Dangwal. The unit carried out detachments at Dewanagri and Chidung to assist Bhutan in airlifting mining equipment. During 1972, it provided extensive transport support to Bangladesh, Bhutan, Sikkim, Orissa, West Bengal and Assam. In 1973, on urgent requisition from King of Bhutan, Triple One was called upon to undertake mercy missions (dropping of food grains) in Bhutan. It flew 179 sorties airlifting a total of approx 8900 kg of food grains. It had the honour of flying sorties for the coronation of King of Bhutan in Jun 1974. On 2nd Sep 1974, there were heavy floods in West Bengal and Bihar. During this period, 45 men, women and children saved by Sqn Ldr MS Dhillon in marginal weather conditions when repeated attempts by army boats had failed to reach these marooned people.

From July to August of 1975, it provided relief to victims of ravaging floods in Orissa and Bihar by dropping 96 tonnes of food grains and rescuing 75 passengers. On 21st Mar 1976, it received the first consignment of the light and agile Chetak helicopters for replacing the old workhorse Mi-4 under the re-equipping scheme; and, in May 1983 it moved to Barielly. In 1986, the unit was designated as an armed helicopter unit and was affiliated to an armoured division. On 18th Aug 1998, when disaster struck Malpa village of Pithoragarh district, the Snow Tigers were back in action to carry 20,605 kg, 119 passengers, 37 dead bodies and 15 casualties in 185 sorties. Currently, the unit is reequipping with ALH and is actively involved in all the relief and rescue missions in the sector covering Kumaon and Garhwal hills. On a regular basis, pilots of this unit operate from sea level to the daunting altitudes of over 17,000 ft.

In Feb 1993, it participated in Desert Arrow with the Indian Army. In Sep 1993, it was involved in massive flood relief operations in Terai Region of Nainital and Bareilly division. In Feb 1995, it participated in Athambore; and in 1996, it took part in Grand Slam with four helicopters and 10 pilots. Helicopters operated from forward helipads with Army under field conditions and flew nearly 268 hours towards anti-tank missions and integrated training under realistic conditions. In the same year, it took part in Little Slam in which 43 hours were flown. In Oct 1996, it took active part in exercise Himalaya, another large joint endeavour. On 11th May 2004, it achieved greater heights by carrying out the highest landing by a Cheetah helicopter at an altitude of 23,250 ft for evacuating mountaineers of Air Force Mount Kamet expedition team, which incidentally is a world record. The participating aircrew, Wg Cdr SK Sharma and Flt Lt AB Dhanake were cited by the Smithsonian Air and Space Museum at Washington DC on 5th Apr 2005 with the Laureates Trophy for using aviation for the betterment of mankind. The only other Indian recipients are Mr JRD Tata and His Excellency, The President of India Dr. APJ Abdul Kalam.

118 HU   In June 1972, the unit undertook communication duties during the Shimla Accord signed between Mrs Indira Gandhi and Mr ZA Bhutto. It carried out extensive flood relief operations in Cachar district of Assam, air dropping 13,774 Kg of food supplies in May 1973. In May 1975, it airlifted emergency food supplies to central Bhutan involving a total of 34,117 kg. In June 1973, the unit commenced air maintenance operations from Chabua, Jorhat and Mohanbari. In April 1974, it undertook communication duties for the Prime Minister of India, other heads of States, Ambassadors and other high-ranking officials for the coronation of King of Bhutan. Emergency food dropping was carried out in Nagaland and Arunanchal Pradesh in Aug 1975 and flood relief and rescue operations in Bihar in September. In Jul 1980, the unit airlifted a total of 109612 kg of food supplies for Mizoram and 63,222 kg of supplies for the flood victims of Orissa. In July 1994, it was called upon to provide succour to the flood victims of Orissa. In a short period of nine days a total of 145,000 kg was airlifted. The then Chief Minister of Orissa, Shri Biju Patnaik personally wrote a letter to the Chief of Air Staff in appreciation of the unit's efforts. In Jun 1998, it was called upon to carry out extensive flood relief operations in Subansiri district of Arunanchal Pradesh.

It carried out special heliborne operations to induct Indian Army troops into Nagaland in Sep 1974, in order to capture Naga hostiles trying to enter India from China; a total of 1000 troops were airlifted. In Oct 1987, two helicopters of the unit took off for Jaffna, Sri Lanka as part of IPKF. The aircraft were extensively employed for troop induction, logistic support and casualty evacuation; for which Wg Cdr CD Upadhyay was awarded the Vir Chakra. In May 1975, it was tasked with ferrying from main land India to Andaman Nicobar Islands and carrying out trial landings in the Andaman Islands. This resulted in the establishment of an inter-island helicopter courier, and subsequently in establishing a Mi-8 flight at Car Nicobar.

116 HU   In Dec 1972, the unit carried out live para dropping trials. A fair amount of accuracy was accomplished. In Dec 1973, Allouette were inducted

and the role of the unit was changed to communication, casevac and anti-tank operations. It participated in the Air Force Fire Power demonstration held in Feb 1978 and the first joint army/ air force exercise in the ATGM role. To cater for the anti-tank operations role, the Aloutte III was modified to carry four AS II BI missiles. In Sep 1978, flood relief operations in Delhi were a highlight. In Mar 1979, it flew for the Govt of Himachal Pradesh in assisting the avalanche stricken people in the Lahaul Spiti Valley. On 1st Nov 1986, the unit came under the command and control of the Indian Army and was affiliated to an armoured brigade under a Corps at Jodhpur. It moved to its location at Jodhpur on 14th Jan 1988. On 27th Jan, it was tasked to operate a two aircraft detachment in Op Pawan (Sri Lanka) in the anti-shipping role. Later, the detachment moved to Jaffna and remained there till Jan 1989. During this period a total of 43 missiles were fired at LTTE hideouts in the jungles. In Jul 1990 the unit carried out flood relief operations, dropping flood and medicines in the area around Jodhpur, Pali and Balotra.

115 HU　A primary role of the helicopter being rescue and saving lives, the unit provides prompt aid to civil authorities in times of natural calamities and disasters not only in the entire North East but wherever required.

In the month of Aug 1975, it took active part in flood relief operations in Bihar to provide much needed relief and succour to marooned and homeless people. It was during these operations that the national media christened this unit as "The Hovering Angles" who came to the rescue of the people in Bihar. Floods struck again in Oct 1978, where it maintained a three aircraft detachment at Calcutta. Thus commenced a major flood relief operation in

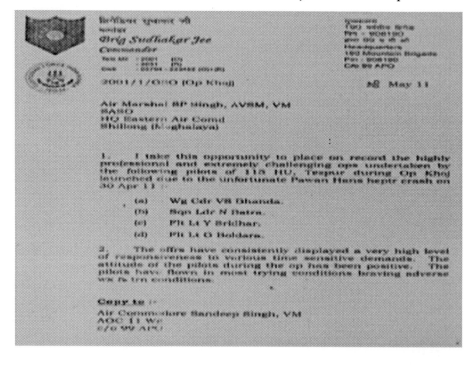

which tireless pilots of the unit flew from dawn to dusk for over two weeks. A total of 296 sorties utilizing 335 hours in a mere 14 days was indeed remarkable and commended by one and all. The unit had in similar ways provided relief in places as far as Andhra Pradesh and Orissa. Cyclone relief operations were carried out in May 1979. The unit flew 23 hrs and dropped 4.5 tons of load towards flood relief operation in Orissa in Sep 1980. In Jan 1987, Arunachal Pradesh experienced some of the worst landslides due incessant rains. 11 landslide victims were picked up from extremely inhospitable and inaccessible terrain at a very short notice.

127 HU Baptized in the lofty peaks of the Himalayas, the unit boasts of a commendable air maintenance tonnage legacy in support of the Indian Army. In Op Falcon in the northeast, the unit holds a whopping unbroken record of over 5500 MT. Primacy has always been given to joint exercises and operations such as Eagle-86, Trishakti, Brass Tacks IV, Op Pawan, Meghdoot and Falcon etc. Inter-service relations were reinforced with the underslung transportation of a crashed AOP Cheetah aircraft from Harike in Mand area near River Beas on 31st Jul 1986, and similarly the underslung trials of medium girder bridges and main light assault bridges for 61 Engr Regt on 18th Aug 1986. The onset of monsoons created havoc in the Kameng Sector in Arunachal in Jun 1989 and a major relief effort was launched by this unit from Tezpur. A huge task seemed to be imminent since the only road from Bhalukpong to Tse La Pass was blocked with about 78 landslides. It airlifted over 600 tonnes for the bereaved population by keeping the flying machines airborne from dawn to dusk for many days.

142 SSS Flight It was formed at Jamnagar on 1st Aug 1974 with Mi-4 helicopters and was entrusted with the task to support forward airbases, visual observation posts and air land / air sea rescue. The basic war time role of the flight is helicopter support in general or as required basis and other special tasks as and when ordered. While at Jamnagar, the flight rendered timely help to the state government when during floods in Gujrat

in 1976, the unit dropped approximately 2300 kg of food packets in the "Little Rann" area. It also rendered relief to the state of Tamil Nadu, Andhra Pradesh and Gujarat in 1979. The flight converted to Chetak helicopter on 1st Sep 1981 and moved to Kalaikunda on 7th May 1982. It again moved to Bagdogra on 6th Dec 1982. It was actively involved in flood relief operations in Aug 1982 at Bhubneswar; dropping 27686 kg of food packets in 173 sorties totalling 190:20 hours of flying, commendable job indeed for a flight of its size. On 7th Oct 88, the Chetak was pushed to its limits in the safe evacuation of foreigners belonging to Austria, Switzerland and Germany from high altitudes in Bhutan. The evacuation was done from an altitude of 14300 ft in extremely critical weather condition with a total flying effort of eight hours in one day. It played an invaluable role in the successful launch of Agni and its subsequent launches by flying a total of 123 hours in 130 sorties for various tasks. It also has the distinction of carrying out the first ever SAR by night (in Chetak) in IAF. This small unit has always stood by to give aid to civil authorities of West Bengal, Bihar and Sikkim.

129HU   This helicopter unit can proudly claim to be a frontline combat arm of the IAF today. It has operated the length and breadth of the land from the icy waste of the Karakorum in the north to the plateau of the Deccan in the south. It has participated in two active operations; one in the highest and most inhospitable battlefield in the history of warfare and other across the seas in thick jungles of strife-torn Sri Lanka. If the unit has operated in a variety of roles in the war, it has not lagged behind its contribution during peace. It provided aid to civil administration in a number of states such as Punjab, Gujarat, Bihar, Himachal Pradesh and J&K. The nature of commitments undertaken have been equally varied, ranging from flood relief in plains, to dropping of relief materials and essential supplies to stranded snow-trapped civilians in mountainous passes, and from counter terrorist operation in Punjab to supporting heli-skiing in Kashmir; and at other end of the scale, flying offshore to the oil rigs and platforms of ONGC in the Arabian Sea.

Apart from regular military tasks and joint operations, the unit regularly engages in providing airlift to VVIP/VIPs of India and foreign countries. It also undertakes various tasks of providing aid to civil power in case of natural calamities like floods, riots, casevac etc. It carried out flood relief operations in Bihar in 1987, in Gujarat, Punjab and HP states in 1988 and at Malpa (Pithoragarh) in Aug 1998. It also became the first Mi-17 unit in the IAF to undertake armament trials at SK Range in Aug 1988. In the month of Jul 1988, apart from the unit's operational commitment in Siachen, six helicopters were dispatched to take part in counter terrorist ops in the Mand area of Punjab. During this operation, 1350 Punjab police personnel were airlifted, utilizing 49:20 hours in 84 sorties. The month of Aug & Sep 88 also witnessed devastating floods in the north Indian states of Punjab, Gujarat and Bihar. The heavily committed unit was again pressed into service in aid to civil authorities, and total of 79/21/34 sorties utilizing 48:20/31:45 / 30:20 hours were flown respectively for Punjab, Gujarat and Bihar. In 1988, it won its first richly deserved laurels when it was awarded the "IAF Squadron of the year" trophy 1988. For a newly raised fledgling unit to be awarded such a prestigious award within a year of its raising, speaks volumes for the sincere hard work and dedication of all ranks.

132 FAC Flight   In Oct 2006, the flight air lifted 1119 kg of load for the Indian Army to organize their communication set up on a difficult terrain at Anganpatri (10500 ft) situated right on the LOC in the Udhampur sector. During Op Imdad in Srinagar sector, the flight flew 410 sorties, 173:40 hours and evacuated over 255 casualties, 100 dead bodies and dropped 125 tons of load. It was involved in rescuing Indian Army personnel trapped in an avalanche in Sonamarg in Feb 2012. By far, the best utilization of the aircraft of this unit has been for SAR and casevac duties. A humble figure of 3200 casualties lifted till date since its inception goes a long way in highlighting the motto Apatsu-Mitram. A majority of the requisitions are for aid to the civil power and for picking up critical cases in difficult mountainous areas. The unit has been lauded for its operations in support of the state of J&K. For example, the case of rejuvenating the power supply of entire Srinagar valley by winching down electricity grid maintenance staff at Banihal pass following a black out in valley after avalanches. Similarly, it rescued nine souls buried in an avalanche during the early part of 2012. The unit remains prepared 24×7 for any contingency.

128 HU: Zoji-La Rescue   This incident made front page newspaper headlines and Wg Cdr Dutt was awarded the coveted Kirti Chakra in recognition of his efforts. The Hindustan Times, in their article published on 29th Nov 1987 called him the 'Hero of Zoji-La' for saving 145 persons trapped in the avalanche and carrying out 54 sorties in four days. The newspaper mentioned, "*It was already 3:30 PM when 'Mike' Dutt landed at Kargil for refuelling with sunset time fast approaching. He had to decide whether to carry on or stay the night. The task was urgent and a delay would mean loss of lives. So he decided to carry on because he sensed that the weather would pack up. The weather did pack up and what followed is history*". What could have been a three hour flight along the Indus River was cut short by half. Dutt with co-pilot, Flt Lt James and two sergeants was in the deep and narrow valley at 5:25 PM flying straight into the setting sun with reducing visibility. The temperature was minus 15 degrees Celsius and the trapped survivors were located with some effort. The newspaper further quoted Wg Cdr Dutt as saying "I did a dummy run and a quick assessment of the terrain and found the snow kicking up and blocking the view, so I decided to land with forward speed. I asked the co-pilot to survey the ground closely and warn if the landing needed to be abandoned." When they landed the co-pilot reported that the chopper was sinking and tilting to the right but now it was too late to abandon the landing. The only alternative was to keep the helicopter on collective and keep it light on wheels. The sergeants rushed to the trapped people. Three of them needed to be helped into the helicopter because of their injuries. Also waiting on the spot were three serious casualties of the Indian Army. Total time taken on ground was 15 minutes.

122 Helicopter Flight   It was established on 23rd September 1981 at Kalaikunda. The flight moved to Carnicobar on 1st April 1985. It is equipped with Mi-8s, a machine that has been the mainstay of the nation's medium lift rotary wing capacity for many years, from the icy frontiers of Siachen Glacier to the jungles of Assam. This helicopter is capable of an astonishing variety of roles – from accurate bombing and rockets attacks, to ferrying large quantities of load; from carrying out rescue and combat SAR to special military missions with the army and the navy.

In Nov 1983, Sikkim suffered one of its worst ever landslides. The flight undertook air maintenance for army and GREF in the landslides relief operations at Gangtok and 75,914 relief bags were airlifted and dropped. It received accolades in abundance for this mission. In Feb 1984, one Indian Navy Chetak helicopter force-landed in the Sunderban jungles. A Court of Inquiry team was airlifted and dropped from a Mi-8 helicopter of this flight to the accident spot by live winching. Later in the month, the force-landed helicopter was transported by Mi-8 as an underslung load. The flight participated in Op Leech at Port Blair from 7th to 10th Feb 1998, wherein a platoon of troops was slithered down at East Island (North Andaman) to counter smugglers. 22:30 hours in 20 sorties was put in to these joint operations. This flight also participated in the Independence Day fly-past during the month of Aug 1998 at Port Blair. The unit has since then done yeoman service for the people and civil authorities of Andaman & Nicobar. Thousands of mercy missions, couriers, anti-piracy effort, search and rescue and communication flights are testimony to the immense contribution.

The Flying Dolphins is the only flying unit of the IAF in entire Andaman & Nicobar region, and has been a very visible symbol of the nation's resolve to stand by its citizens through thick and thin; never more than the aftermath of the deadly tsunami in December 2004. During this terrible event, the airmen and officers of the unit swung into action even as the gigantic waves were still crashing with full ferocity on the islands. Unmindful of their own families, the air warriors began rescue work like a well-oiled machine. They managed to rescue 350 persons on the fateful day with their ceaseless flying effort and dauntless spirit, winching-up people from the most difficult and dangerous locations. As part of the tri-services Andaman and Nicobar Command, it shoulders a number of important responsibilities in peace and war. Since its move to the present location, the flight has undertaken many operational missions that have been instrumental in securing peace and security in this far-flung frontier of the nation. The flight is central to the anti-poaching and anti-piracy drive in the Andaman Sea. A very important part of the unit responsibilities is Search and Rescue over the high seas – a role in which the Flying Dolphins have seen much action, and won fulsome praise of the country.

It was instrumental in rescuing a large number of service and civil personnel at Carnicobar in the wake of the deadly Tsunami on 26th Dec 2004. The flying Dolphins regularly provide succour to ships in distress and stranded survivors. Be it the evacuation of scientists from Barren Islands on 10th July 2004, or the rescue of navy personnel in South Bay in Jul 2006, or the rescue of eight survivors from MS Soni, a vessel capsized off the North Brother Island in Jan 2007, or the daring rescue operation for INS Mahish in Jan 2010; this flight has been at the centre of all relief missions in the A&N area. The flight has given a good account not only of itself, but also of the country's through some famous and high visibility relief missions. One such operation was the rescue of nine crew members of the Malaysian tug Sea-Mariam near Port Blair in 2006. Another famous rescue involved 14 crew members of Indonesian MV MAC, a 7000 tonne behemoth in Aug 2006. On 22nd May 2011, a boat capsized off Carnic, in which a number of tribals were feared missing. Upon urgent request from the Carnicobar Deputy Commissioner, the Flight launched an SAR mission at 2100 hours, within 30 minutes of requisitioning. The search was continued early next day, amounting to three hours of SAR flying. On 27th Jul 2011, the Flight launched an ASSR mission to provide succour to the police detachment at Tillanchang where a drowning death had occurred

The Unit has contributed fully towards fostering true jointmanship by active participation in Tri-Services events. It has remained at the forefront of joint exercises, having regularly conducted joint training with Indian Army, Special Operations camps with MARCOS, as well as exercises with the Indian Navy. Inter-island night flying and Live Helocasting training with MARCOS are regularly done.

152 HU   The Mighty Armour has proved to be the lifeline of the troops in the Kashmir Valley and elsewhere, ferrying endless passengers and supplying rations, ammunition and much needed relief in the harsh climes of the beautiful valley. Peace time roles include route transport role, air logistics, casualty evacuation, miscellaneous tasks, aid to civil authorities, battle indoctrination, VIP/VVIP tasks, and counter-insurgency operations. A snapshot of multifarious tasks undertaken, other than those dedicated to routine air maintenance and joint exercises with the Indian Army, shows the range of capabilities an IAF helicopter unit possesses.

Sl No	Period	Agency	Description
1	Oct 88	Flood Relief (Govt of Punjab)	Soon after the formation of the unit in Sep 88, the unit was tasked to undertake flood relief operations in Punjab area. The unit flew 131 hours, airlifted 91 tonnes of load and 344 casualties.
2	Dec 88	Flood Relief (Govt of HP)	Carried out flood relief operations in Himachal Pradesh. Flew 254 hours and airlifted 5821 passengers.
3	Oct 91	Flood Relief (Govt of Punjab & Haryana)	Carried out flood relief operations in Punjab and Haryana.
4	Apr 92	Search & Rescue	There was a mid-air collision of two AN-32s on 01 Apr 92. It undertook SAR and was airborne in a record time of 20 min. In that dark night, the wreckage was spotted after adopting non-conventional search methods. The helicopter was landed next to the crash site in an area which was interspersed with high tension cables and trees, to render assistance to the fellow servicemen.
5	Aug 92	Casualty Evacuation	On 05 Aug 92, it was tasked to airlift rations for the inhabitants of Harsil area who were cut off due to a landslide. Airlifted 18 tonnes and evacuated 262 stranded pilgrims and army personnel in two days.
6	Sep 95	Flood Relief (Govt of HP, Punjab & Haryana)	Undertook flood relief operations in the states of Punjab, Haryana and Himachal Pradesh.
7	Sep 97	Casualty Evacuation	Undertook a challenging dark night sortie of evacuating 16 critically injured NSG troops from Qazigund to Srinagar.
8	Jun 98	Casualty Evacuation	A total of 11 IMA cadets were evacuated from Gomukh at Gangotri glacier after being hit by an avalanche.
9	Aug 98	Landslide & Flood Relief (Govt of UP)	Operated for landslide and flood relief operations from Pithoragarh and Gorakhpur in Malpa area. It flew 13 sorties clocking 12 hours and airlifted 940 kg and 70 passengers.

Sl No	Period	Agency	Description
10	Nov 98	**Disaster Relief (Govt of UP)**	Flew eight sorties clocking five hours and airlifted 13 tonnes & 74 passengers towards landslide at Joshimath area.
11	Mar 99	**Earthquake Relief (Govt of HP)**	Flew 56 hours when earthquake hit Chamoli region and airlifted 63 tonnes of load.
12	Aug 2000	**Flood Relief (Govt of HP)**	Carried out relief operations during flash floods in Satluj river.
13	Oct 2000	**Mi-8 Salvage (Rann of Kutch)**	Carried out 53 live winching sorties clocking 21 hours over two days in Kutch area to recover the wreckage of Mi-8 helicopter from the crash site.
14	Jan 01	**Earthquake Relief (Govt of Gujarat)**	Carried out earthquake relief operations in Bhuj at Gujarat. The unit air dropped 63 tonnes of load.
15	Nov 01	**Casualty Evacuation**	The unit carried out evacuation of 16 persons who were stranded at Chandra valley near Kunzum La pass from an altitude of 14000 ft.
16	Dec 03	**Casualty Evacuation (ITBP)**	Carried out evacuation of ITBP personnel in nine sorties clocking five hours.
17	Sep 03	**Offshore operations (ONGC)**	Offshore operations were restarted after seven years by IAF. It operated from Sep 03 to Dec 03 and flew 169 sorties clocking 121 hours and airlifted seven tonnes of load and 2698 passengers.
18	Aug 04	**Casualty Evacuation (Govt of Uttarakhand)**	The unit carried out evacuation of 348 pilgrims from Badrinath in 56 sorties clocking 19 hours.
		Flood Relief (Govt of Uttarakhand)	Carried out relief operations in Uttarakhand in 12 sorties clocking seven hours and lifted 19 passengers.
19	Oct 04	**Casualty Evacuation (Govt of Uttarakhand)**	Carried out evacuation missions in 12 sorties clocking seven hours and evacuated 16 patients.
20	Nov 04	**Casualty Evacuation (Govt of Uttarakhand)**	Carried out evacuation missions from Gopeshwar to Joshimath in six sorties clocking five hours and evacuated 30 pilgrims.
21	Jul 04	**Flood Relief (Govt of Bihar)**	Carried out casualty evacuation and supply dropping missions after the floods in Bihar. The unit flew 150 sorties clocking 146 hours and airlifted two tonnes of load and 424 passengers.
22	Nov 04	**Casualty Evacuation (Army)**	Carried out evacuation of five casualties.

Sl No	Period	Agency	Description
23	Dec 04	**Casualty Evacuation (Army)**	Carried out evacuation of three casualties.
24	Jan 05	**Casualty Evacuation (Army)**	Carried out evacuation of one casualty.
25	Feb 05	**Anti-Naxal Surveillance (Govt of Bihar)**	Carried out surveillance missions in Bihar area and airlifted 277 passengers.
		Anti-Naxal Surveillance (Govt of Jharkhand)	Carried out surveillance missions in Jharkhand area and airlifted 440 kg of load, 823 passengers and one dead body.
26	Mar 05	**Casualty Evacuation (Army)**	Carried out evacuation of five casualties.
		Casualty Evacuation (Govt of UP)	Carried out evacuation of 12 casualties and one dead body.
27	Jul 05	**Flood Relief (Govt of HP)**	Carried out relief operations at Rampur area of Himachal Pradesh after the floods. A total of 174 sorties were flown clocking 98 hours and airlifted 78 tonnes of load and 1282 passengers.
		Casualty Evacuation (Govt of Uttarakhand)	Flew three sorties clocking two hours and airlifted six casualties from Uttarkashi to Dehradun.
		Casualty Evacuation (ITBP)	Undertook evacuation of nine ITBP personnel.
28	Jul – Sep 05	**Casualty Evacuation (Taklakot)**	Undertook six sorties clocking six hours for evacuating one casualty from Taklakot in Chinese territory to Bareilly.
29	Jul – Sep 05	**Casualty Evacuation (Army)**	Flew 10 sorties clocking five hours airlifting four casualties from Roorkee, Rampur and Srinagar.
30	Sep 05	**Casualty Evacuation (Ministry of External Affairs)**	Flew six sorties clocking seven hours evacuating 11 casualties (Italian nationals) and one dead body.
		Earthquake Relief (Govt of J&K)	Carried out relief operations at Uri sector of J&K. Flew a total of 255 sorties clocking 110 hours and airlifted 96 tonnes and 756 passengers.
31	Oct – Dec 05	**Casualty Evacuation (Army)**	Flew six sorties clocking four hours evacuating four casualties from Dehradun to Roorkee.
32	Jan – Mar 06	**Casualty Evacuation (Army)**	Flew three sorties clocking two hours towards casualty evacuation of Army personnel.
33	Apr – Jun 06	**Casualty Evacuation (Army)**	Flew 10 sorties clocking eight hours and evacuated six casualties from Karcham to Roorkee.

Sl No	Period	Agency	Description
34	Aug 06	**Flood Relief (Govt of MP)**	Carried out relief operations from Bhopal towards the flood in the area. The unit flew a total of 19 sorties clocking 12 hours and airlifted 135 survivors and winched up two persons.
35	Jul – Sep 06	**Casualty Evacuation (Army)**	Carried out eight sorties clocking seven hours and evacuated six casualties from Karcham to Roorkee.
36	Oct – Dec 06	**Casualty Evacuation (Army)**	Flew eight sorties clocking seven hours and evacuated six casualties from Roorkee to Dehradun.
		Casualty Evacuation (Govt of Uttarakhand)	Evacuated seven school children with burn injuries from Haldwani to Safdarjung in three sorties clocking three hours.
		Casualty Evacuation (Govt of J&K)	Carried out casualty evacuation missions in seven sorties clocking six hours.
		Casualty Evacuation (ITBP)	Undertook evacuation of two ITBP personnel from Sumdoh in three sorties clocking four hours.
		Air Logistics (DGBR)	The unit flew 10 sorties clocking seven hours and lifted eight tonnes of load.
37	Dec 06	**Casualty Evacuation**	The unit immediately evacuated 15 school children during Vijay Diwas events from Hindan in four sorties clocking three hours.
38	Jan – Mar 07	**Casualty Evacuation (Army)**	Flew seven sorties clocking four hours and evacuated two casualties from Rajouri and Roorkee.
		Air Logistics (DGBR)	The unit flew 30 sorties clocking 21 hours and lifted four tonnes of load and 120 passengers.
		Air Logistics (Govt of J&K)	Flew 16 sorties clocking eight hours and lifted 42 passengers.
39	Apr – Jun 07	**Air Logistics (Govt of J&K)**	Flew eight sorties clocking three hours and lifted 92 passengers.
40	Jul – Sep 07	**Flood Relief (Govt of Bihar)**	Flew 25 sorties clocking 37 hours towards the relief operations at Bihar.
41	Jan – Mar 08	**Air Logistics (Govt of J&K)**	Flew 48 sorties clocking 25 hours and lifted 476 passengers.
42	Jul – Sep 08	**Casualty Evacuation (Govt of HP)**	Flew 15 sorties clocking 15 hours and evacuated 39 casualties and airlifted two tonnes of load.
		Casualty Evacuation (Ministry of External Affairs)	Flew two sorties clocking two hours and evacuated eight casualties (Belgian nationals). The unit also flew three sorties clocking four

Sl No	Period	Agency	Description
			hours and evacuated 11 casualties (Israeli nationals).
43	Jan 09	**Casualty Evacuation (Govt of HP)**	Relief operations in 17 sorties clocking 13 hours and airlifted 275 passengers.
		Air Logistics (Army)	62 sorties clocking 26 hours and lifted four tonnes of load and 228 passengers from Bimbat sector.
		Air Logistics (Govt of J&K)	Nine sorties clocking seven hours and lifted 63 passengers.
		Air Logistics (DGBR)	12 sorties from Kullu and Kargil clocking seven hours and lifted 650 kg of load and 36 passengers.
44	Feb 09	**Casualty Evacuation (Army)**	Evacuation missions from Chakrata to Delhi in three sorties clocking three hours towards Army.
		Air Logistics (Army)	21 sorties clocking 10 hours and lifted three tonnes of load and 116 passengers.
		Air Logistics (Govt of J&K)	28 sorties clocking 20 hours and lifted 16 tonnes of load and 56 passengers.
		Air Logistics (DGBR)	15 sorties from Kargil clocking 12 hours and lifted one tonne of load and 32 passengers.
45	Mar 09	**Air Logistics (DGBR)**	25 sorties from Kargil clocking 14 hours and lifted six tonnes of load and 28 passengers.
46	May 09	**Air Logistics (DGBR)**	Four sorties from Kargil clocking five hours and lifted 300 kg of load and 22 passengers.
		Election Duties (Govt of J&K)	10 sorties clocking 11 hours and lifted 42 passengers towards the election in J&K.
		Air Logistics (Army)	Air logistics from Awantipur. The unit flew 16 sorties clocking 10 hours and lifting 61 passengers.
		Casualty Evacuation (Army)	Unit flew one sortie clocking one hour evacuating casualties from Halwara to Chandimandir.
47	Jun 09	**Air Logistics (DGBR)**	35 sorties from Dharchula clocking 21 hours and lifted 30 tonnes of load and 13 passengers.
48	Jul 09	**Air Logistics (DGBR)**	16 sorties from Dharchula clocking 11 hours and lifted eight tonnes of load.
49	Nov & Dec 09	**Election Duties (Govt of Jharkhand)**	The unit operated a four helicopter detachment at Jharkhand for election duties. A total of 475 sorties were utilized for airlifting 16 tonnes of load and 3741 passengers. Apart from the main role of airlifting polling parties, the unit carried out numerous casualty evacuation and area domination sorties in Naxal prone areas.

Sl No	Period	Agency	Description
50	**Dec 09**	**Air Logistics (GREF)**	Air logistics missions from Sase and Stingri. Flew a total of eight sorties clocking four hours and lifted 13 passengers and three tonnes of load.
51	**Jan 10**	**Air Logistics (GREF)**	Air logistics missions from Sase and Stingri. Flew a total of 15 sorties clocking seven hours and lifted 29 passengers and five tonnes of load.
52	**Feb 10**	**Casualty Evacuation (MHA)**	Carried out evacuation missions from Jagdalpur. Flew two sorties clocking three hours.
		Casualty Evacuation (DRDO)	Carried out evacuation missions from Jagdalpur. Flew two sorties clocking two hours.
		Casualty Evacuation (Govt of J&K)	Carried out evacuation missions from Bimbat. Flew three sorties clocking two hours.
		Air Logistics (GREF)	Air logistics missions from Kargil. Flew a total of six sorties clocking four hours and lifted seven passengers and 300 kg of load.
		Air Logistics (Army)	Air logistics missions from Kargil. Flew a total of 62 sorties clocking 33 hours and lifted 27 passengers and four tonnes of load.
53	**Mar 10**	**Air Logistics (GREF)**	Air logistics missions from Sase and Stingri. Flew a total of 18 sorties clocking 11 hours and lifted 38 passengers and three tonnes of load.
		Air Logistics (BSF)	Air logistics missions from Raipur. Flew a total of two sorties clocking one hour and lifted nine passengers.
		Air Logistics (Army)	Air logistics missions from Awantipur. Flew a total of 34 sorties clocking 17 hours and lifted 51 passengers and four tonnes of load.
54	**Apr 10**	**Air Logistics (GREF)**	16 sorties clocking 10 hours and lifted three tonnes of load.
55	**Jul 10**	**Air Logistics (Army)**	14 air logistics sorties for Army from Srinagar, clocking seven hours and lifted 10 tonnes of load.
56	**Nov 10**	**Air Logistics (Army)**	One sortie clocking two hours in Jagdalpur area.
57	**Jan 11**	**Air Logistics (GREF)**	14 sorties clocking seven hours and lifted 20 passengers and two tonnes of load from Bhuntar.
		Air Logistics (Army)	58 sorties clocking 25 hours and lifted 20 tonnes of load and 69 passengers in Srinagar valley.

Sl No	Period	Agency	Description
		Air Logistics (Govt of J&K)	11 sorties clocking five hours and lifted 92 passengers.
58	Feb 11	Air Logistics (BRTF)	15 sorties clocking eight hours from Bhuntar. A total of two tonnes of load and 32 passengers were lifted.
59	Mar 11	Air Logistics (BRTF)	Air logistics missions from Sae and Stingri. Flew 13 sorties clocking seven hours and lifted 35 passengers and two tonnes of load.
60	Apr 11	Air Logistics (GREF & BRTF)	17 sorties clocking nine hours from Stingri and Sase and lifted 300 passengers and three tonnes of load.
		Air Logistics (Govt of HP)	16 sorties clocking 13 hours and lifted 216 passengers.
61	May 11	Election Duties (Govt of West Bengal)	16 sorties clocking nine hours and airlifted 63 officials.
62	Jun 11	Air Logistics (Army)	Two sorties clocking one hour in Chhattisgarh area.
63	Dec 11	Air Logistics (Army)	45 sorties clocking 21 hours lifting 12 tonnes of load and 61 passengers in Srinagar valley.
		Air Logistics (GREF)	11 sorties clocking eight hours and lifted three tonnes of load and 18 passengers in Stingri and Sase.

IN SERVICE TO OUR COUNTRYMEN: GALLANTRY AWARD CITATIONS

Kirti Chakra

Wing Commander Maheshwar Dutt (11607), Flying (Pilot)

On the evening of 14th November 1986, an expected and very high snowfall blocked both ends of the 3400 meter high Zozi-la pass. Over one hundred vehicles were trapped in the pass along with their occupants. Numerous landslides and avalanches further added to the disaster. Wing Commander Maheshwar Dutt, flying a Mi-17 helicopter, flew a total of 75 rescue sorties, 56 of them in the first four days. He personally rescued 145 persons saving them from almost certain death in the raging blizzard with wind speeds of 70 kmph and temperature well below zero. On no occasion was a prepared helipad provided and Wing Commander Dutt often landed on compact snow by the roadside. These operations, from makeshift helipads scarcely larger than his helicopter wheel base were carried until all the trapped persons were evacuated.

Wing Commander Sudhir Kumar Sharma (17717) Flying (Pilot)

On 11th May 2004, Wing Commander Sharma was directed to urgently evacuate three members of a mountaineering expedition to Mt Kamet, who were critically injured due to inclement weather. These casualties were at an altitude of 23,260 ft, which is beyond the service ceiling of 23,000 ft of the Cheetah helicopter. Due to their serious injuries, it was impossible to bring them down. He flew in his Cheetah helicopter to the site and on locating the survivors realized that the only option was to land on an unprepared small tabletop sloping area located on an exposed snow covered ridge. Such landing demanded extraordinary courage and exceptional flying skills. Even a slight mishandling of helicopter controls could lead to a catastrophic accident endangering the life of the pilot.

No one in the world had ever attempted landing a helicopter at such a high altitude let alone evacuates casualties from an unprepared, snow bound and sloping piece of ground. Despite the deterring and trying conditions, deteriorating weather, severe turbulence, gusty jet stream winds exceeding 120 kmph and low margin of power available, he displaying determination, utmost courage and an utter disregard for his personal safety, landed at the site not once but three times to evacuate the casualties. His courageous act under these near impossible conditions was instrumental in saving the lives of three mountaineers from certain death. This rescue from an altitude of 23,260 ft is the highest ever landing and casualty evacuation by any helicopter and in the process set world record which has been recognized by Limca Book of World Records.

Shaurya Chakra

Sqn Ldr Pranay Kumar (25603) Flying (Pilot)

On 2nd Sep 2007, he was detailed as an additional pilot for a rescue mission. The mission entailed winching up of a citizen stranded in the middle of a gorge of Chambal River affected by flash floods. The prevailing conditions of total darkness were not conducive for operations with unaided vision. The crew took a decision to proceed with the mission on Night Vision Goggles (NVGs) which provided them with better situational awareness. Even though he was an additional pilot seated in the cargo compartment, he donned his NVGs and helped the pilots in locating the survivor and also in giving clearance from obstructions around the vicinity of the survivor.

When the winch cradle was lowered, the survivor could not strap himself to the winch due to extreme fatigue caused by battling the water currents for over six hours. At this juncture, Sqn Ldr Pranay Kumar immediately volunteered to go down to rescue the survivor. Knowing full well the peril to his life, he briefed the Flight Gunners on emergency signals and got lowered in the gorge. He guided the crew of the helicopter using hand signals and despite the darkness and strong water current, managed to reach the survivor. On reaching the survivor he realized that the space was insufficient to get down for strapping the survivor to the harness. He instantly took the decision to physically tie the survivor to himself. Having done this while being half immersed in the strong flow of the water, he gripped the survivor and holding onto him tightly, signalled the Flight Gunner to winch them up. Using all the physical strength that he could muster, he held onto the survivor till the end of the operation. In spite of being aware of the danger to his own life, he undertook the mission of saving a fellow countryman without a second thought.

Vayu Sena Medal (Gallantry)

Wing Commander Rajesh Isser (16968) Flying (Pilot)

Wing Commander Rajesh Isser (16968) Flying Pilot is commanding a helicopter unit in the Easter Sector since 17th May 1999. On 11th Jun 2000, East Siang district was inundated by flash floods causing wide spread devastation. Three helicopters were launched for the rescue task and Wg Cdr Isser was to coordinate these ops at Passighat. He personally flew numerous missions and was able to evacuate 46 people from rooftop, treetop, small eroding islands and a capsized boat by rope ladders and winching. This involved hovering over the swirling and ferocious Siang River in poor visibility conditions and heavy rain. A total of 76 people were rescued from the swirling water. Thereafter, he searched for a capsized ferry, which was finally located after an hour of search. The ferry and its 98 occupants were stuck in the middle of the river, with some of the passengers stuck waist deep in quick sand. He was personally responsible for rescuing 58 people, while others were rescued by the second helicopter pooled in for the task.

With his untiring efforts he also coordinated relief operations in East Siang, Yinkiong and Annini district. A total of 2800 marooned people on either side of the Siang were rehabilitated to their homes between 12th Jun 2000 and 26th Jun 2000. He personally flew more than 70 sorties to evacuate more than 1600 persons cut off from their homes entailing operations from river beds, hill side and unprepared areas. On 24th Jun 2000 some survivors were spotted from a helicopter on a cut off island. Wg Cdr R Isser took off within minutes of the SOS call and winched up 47 people to safety from the swirling waters of the Deopani River. This involved flying in a narrow valley, inclement weather and poor light conditions in late evening.

Flt Lt P Upadhaya (18581) Flying (Pilot)

On 13th Oct 1992, the hauling cable of a cable car carrying passengers to 'Timber Trail' holiday resort had snapped and the cable car with ten passengers on board got stranded in the middle of the valley. The Indian Air Force was given the task to rescue the passengers stranded in the cable car which was dangling at a height of 1500 ft above the river bed. As the rescue mission could not be undertaken after sunset, it was launched next morning.

The initial attempt was made by a Chetak helicopter. But it was found to be unsuitable due to inadequate length of winching cable and was replaced by a Mi-17 helicopter. Flt Lt P Upadhaya was tasked to undertake this hazardous mission. It was concluded that the only way the tourists could be evacuated was by winching them up from the cable car. For this, the helicopter had to hover precariously close to the set of cables that ran above the cable car. The winching operation and the hovering helicopter had to be manoeuvred deftly in order to raise those rescued from the cable car through a maze of twisted cables that were holding the stranded cable car suspended in the air. The mission was fraught with danger; the slightest mistake on the part of any team member could end in disaster. The team was required to operate in unison and at their peak efficiency displaying highest level of flying skill, indomitable courage and cool nerves.

The Mi-17 helicopter was manoeuvred to position itself above the cable car; the roof of the car had a dimension of approx 2 ft × 2 ft. At that height, there was also no visual reference for guidance for steady hover and position keeping above the cable car. Turbulence and strong winds made precision hovering even more difficult and hazardous. To complicate matters, the winch of the Mi-17 helicopter had to be accurately passed through the narrow opening of the cable surrounding the cable car in order to lower Major Cresto of 1 Para Commando, on to the top of the cable car. Five persons were rescued in this manner till night set in and rescue operation had to be stopped. The remaining five passengers were rescued the next morning. Even under ideal conditions, hovering is a tiring exercise and difficult to do over long periods. To do this under the adverse conditions for twenty minutes at a time repetitively ten times for each passenger involved exceptional courage and professional skills as also team work. This rescue operation was perhaps the first of its kind attempt in the history of rescue operations by helicopters anywhere in the world.

Sqn Ldr Ashok Kumar (24900) Flying (Pilot)

On 30th Sep 2009 at around 1530 hrs, HTS was tasked to rescue five youths stranded in the middle of Dindi River (120 Km South of Hyderabad) due to flash floods. On receipt of this information, despite his extremely limited experience on type, Sqn Ldr Ashok volunteered for the task. On reaching the site, the crew realized the extreme danger the stranded persons faced due the fast rising current, and swung into action immediately. Sqn Ldr Ashok as winch operator, along with the other crew members, rescued the five stranded persons in an expeditious manner. Any delay in the rescue would have resulted in certain death of these stranded youths.

He again volunteered to take part in the subsequent relief operations conducted by the school in the districts of Kurnool and Mehboobnagar during the floods which devastated Andhra Pradesh and Karnataka in the first week of Oct 2009. During these operations, he was instrumental in rescuing 19 persons marooned in flood waters at different locations around Kurnool as a winch operator. One rescue mission carried out on 5th Oct 2009 was achieved under exceptionally demanding circumstances, wherein the crew had to rescue a family of four members including two small children, who had been marooned on a tree submerged in several feet of fast flowing flood waters for three days. The rescue mission was demanding since the crew could neither see the persons nor access them conventionally. With his ingenuity and quick thinking, he innovatively floated the rescue harness to the marooned family and winched them up under the thick branches of the tree, at times even dragging them for a few feet through the flood waters. To achieve this successful rescue, he displayed a high level of skill, quick and decisive thinking as well as crew co-ordination and control of a high order, as the aircraft had to be continually repositioned in the process of rescue. During the entire flood relief operations he flew numerous sorties from first light to sunset, which included six rescue missions saving 24 lives in highly demanding situations, exacting terrain and marginal weather.

Wing Commander Amitabh Sharma VM (21015) Flying (Pilot)

Wing Commander Amitabh Sharma VM (21015) F (P) is commanding a Mi-17 IV unit in the South Western sector since 03 Mar 08. On 26th Nov 2008, a series of ten coordinated terrorist attacks across Mumbai city exposed India to the first ever urban war like situation. By midnight the terrorist captured many places including Nariman House. On 27th Nov at 2130 hours, he was informed of an early take off the next day for slithering operations over Nariman House. The mission involved slithering down 30 NSG commandos on the terrace of the building inside which terrorists were holding Israeli nationals as hostages. He immediately got in touch with leader of NSG commandos and thereafter meticulously planned the mission for early morning drop on 28th Nov. On reaching the launch pad at INS Kunjali fresh inputs on the situation were taken from NSG commandos on the spot. With helicopter in startup conditions he quickly carried out mission briefing clearly spelling out each crew duties and contingencies. He along with NSG team leader then climbed on top of Kunjali ATC to identify the exact location of Nariman House. After identifying the same, he quickly took off for the mission and followed tactical routing to mislead the terrorists as well as media. On reaching, a quick assessment was made. The roof top was small area of 15 feet × 15 feet and was surrounded by obstructions like high voltage cables, water tanks, dish antenna, tin shades etc. Continuous firing and explosions of grenades could be heard amidst the helicopter noise. The hover was required to be done in such a way that the nose of the helicopter was outside the roof top to ensure safe slithering. A total of 30 troops were slithered down in three quick sorties. The mission displayed highest level of synergy between air and ground forces. The route and hover direction were varied each time to maintain element of surprise.

Flight Lieutenant Yogendra Singh Tomar (28497) Flying (Pilot)

On 14th Nov 2008, he was the co-pilot in the helicopter for picking up three Border Security Force casualties, electronic voting machines (EVMs) and election commission officials. Despite reports of anti-national's activities in the area, he greatly contributed in the planning and execution of a successful landing at the helipad. After picking up the personnel and equipment, during takeoff, the helicopter came under heavy small arms and light machine gun fire. Even after the flight engineer, Sergeant Mustafa Ali was fatally injured and the helicopter sustaining heavy damage to the main rotor blades, fuel tank and tail boom, he displayed courage and composure under fire. The loss of the flight engineer apart from being an important personal loss is a great setback in flying the helicopter even under normal circumstances. With the helicopter badly damaged, the loss of flight engineer is accentuated as he monitors the health of helicopter and operates levers and switches not directly related to flying. Normally, the co-pilot is given the responsibility of navigation. Flt Lt Tomar displayed courage and composure in overcoming grief and shock and assisted the captain in safely flying the helicopter by taking on the added duties of the flight engineer and simultaneously carrying out accurate navigation by the dusk and night to get the aircraft back to safety at Jagdalpur.

Flight Lieutenant Mayanglambam Manimohan Singh (24229) F(P)

On 21st September 2003, our troops on the LoC in the Gurez sector noticed an infiltration attempt by a band of insurgents. There was an immediate need to encircle the insurgents and seal the escape route before they could mingle with the general populace. The insurgents were heavily armed and known to be equipped with machine guns and even the presence of shoulder fired missiles was not ruled out. The hostile terrain precluded any movement of troops on foot in the required time frame. HQ 15 Corps projected a requirement of undertaking the induction of troops to cut off the escape route for the insurgents. The general terrain in the objective area was heavily forested. Flt Lt MM Singh was detailed to carry out troop induction. He planned the mission in a most professional manner and executed it with meticulous precision. The mission required troop induction at an unprepared and un-recced area in a hostile environment. Flt Lt MM Singh, realizing the grave and imminent danger and requirement of extremely high skills, decided to lead the formation of two helicopters. He flew the mission in extremely turbulent wind conditions and deteriorating weather. The landing site, which was near the line of control, had no protection and there was a constant threat of fire from the terrorists, who were hiding in the vicinity. Despite the lurking danger, he kept the aim of the mission uppermost in his mind and displayed exceptional courage by carrying out missions at an un-surveyed site on a ridge at an elevation of 13,000 feet on the first day and four more missions the next day in similar conditions. The successful air operations greatly aided the Indian Army in their anti-terrorist operation.

Flight Lieutenant Rajeev Kumar Chouhan (18556) Flying (Pilot)

On 11st Feb 1992, Flt Lt Chouhan was detailed to undertake air maintenance sorties from MON to various DZs of Assam Rifles in Nagaland Area. During the course of air maintenance after carrying out six uneventful sorties and in the seventh sortie from MON to Chaklanyshu, on giving the command to drop the load over the DZ, he suddenly experienced severe vibrations and juddering followed by partial loss of control of the helicopters. The Flight Gunner immediately warned Flt Lt Chouhan that one of the parachutes had separated from the load and on deployment, had entangled in the tail skid. Sensing the emergency and seriousness of the situation, he kept cool and controlled the helicopter. Maintaining his presence of mind decided to put the aircraft down on the helipad PESAO in the immediate vicinity at an altitude of 1750 meters. During the course of landing on the helipad the parachute cords were further entangled in the tail rotor due to the down wash of the Main rotor resulting in shearing off one of the boxes of main rotor and tail rotor blade, at the same time, damaging the tail gear box to a very great extent resulting in further loss of control and directions of helicopter.

Flying Officer Jyoti Prakash Mathur (11366) Flying Branch (Pilot)

Flying Officer Jyoti Prakash Mathur (11366) of the Flying Branch has been serving with a helicopter unit since April 1968. On 27th April 1971, while serving with the Indian Air Force Detachment Ceylon he was detailed to undertake an operational mission from Colombo to Puttalam and Anuradhapura. On the return flight from Anuradhapura, with passengers on board, his track lay over thick forest and hilly terrain. After flying for 30 minutes the engine of the helicopter failed due to a mechanical defect. He immediately put the aircraft into autorotation, and warned the passengers to take up crash positions and then proceeded to carry out a forced landing. He selected a paddy field on the edge of the forest and skilfully manoeuvred the helicopter to a safe landing.

MI-26: The Behemoth in Action

CHAPTER X

UN Peacekeeping: Partnership with Indian Army

Indian Air Force Contingents (IAC) in UN Missions truly stood tall among all others in the business of peace. Many may not publically acknowledge for many reasons, but deep inside know that the will and professionalism displayed by the Indian Air Force air warriors was second to none. Even the Indian Army, deployed since decades, has gone through its ups and downs, including scathing criticism at times. But report after report, UN-sponsored or independent, concedes that even before 'robustness' became fashionable with the publication and debate surrounding documents such as 'Capstone' and 'New Horizons', none ever found IACs wanting in robustness. This speaks very highly of the Indian Air Force in terms of ethos, ethics and air warrior-like qualities. India has been one of the staunchest supporters of the United Nations and the multilateralism it represents. No other single idea or organisation has contributed more to reducing conflicts in the world in the last six decades. The most visible face of the UN at the grassroots level is its peace building endeavours; and a major pre-requisite and constituent of which is the entire gamut of peacekeeping.

MI-35s in Action during Indian Army Exercises

Engage to Destroy & Deter

Action as a Team or Stand-Alone

Prowlers: Seek to Kill

All Indian Army deployments in the Congo, Sierra Leone, Somalia and Sudan have documented and acknowledged the critical role that IAF helicopters have played in their success, and very often in their very survival against some of the toughest odds. In fact, if there is a role-model of jointness, it is these deployments. Two case-studies, i.e. DRC and Sierra Leone under UN Chapter VII are documented. While Op Khukri in UNAMSIL was a one-off operation that required the finest integration of air and ground capabilities, MONUC (DRC) is about a sustained eight-year commitment to the Indian Army and India's national interest. The study of these two co-located deployments of the IAF and Indian Army, and the extremely successful strike-rates, prove beyond doubt that the Indian model of integration works well where the intent is clear and honest.

Sierra Leone: Op Khukri

Genesis of the Crisis   The trouble and unrest in Sierra Leone dates back to early nineties. After a decade of misrule characterized by corruption, favouritism and mismanagement, some units of the army marched to Freetown in 1992 to protest against the poor conditions of the armed forces. This mutiny quickly took shape of a coup and a military government came to power. Since then, the history of Sierra Leone is replete with military coups, struggles for power, insurgency, instability and anarchy. At the same time Revolutionary United Front (RUF) a fledging organization in the early nineties grew into a major rebel force taking advantage of unstable conditions in the country and help from certain neighbouring countries. By 1995, RUF had established control over major diamond mining areas and was only a few miles away from Freetown. In 1996, a peace accord was signed at Abidjan between the Govt of Sierra Leone and RUF. But peace did not return to Sierra Leone and the RUF continued its attacks.

In 1999, the signing of the UN-brokered Lome accord between President Kabbah of Sierra Leone and RUF leader, Foday Sankoh, led to the birth of United Nations Mission in Sierra Leone (UNAMSIL). This mission comprising troops contributed from eight member countries and military observers from some others was mandated to oversee the process of Disarmament, Demobilisation and Reintegration (DDR) in the trouble-torn country. As part of the Indian Contingent, 5/8 Gorkha rifles, (INDBATT-1), moved to Sierra Leone in Dec 1999. The battalion was deployed at Daru, a small town in the Eastern Province, about 260 Km east of Freetown in Mar 2000. Two companies of the battalion were subsequently tasked to move to Kailahun, a RUF stronghold in Apr 2000. The situation in Sierra Leone deteriorated on 01 May 2000 with the clashes between Kenyan peacekeepers

(KENBATT) and the RUF in Makeni/Magburaka, in the Northern Province, over the issue of ex RUF combatants who had joined the DDR process. The fallout of incidents in the Northern Province resulted in the detention of Indian peacekeepers in Kailahun and Kuiva on 2nd May 2000. Diplomatic and international pressure brought upon the RUF led to release of Indian Peacekeepers. However, the standoff at Kailahun between RUF and the two Indian companies continued, with the RUF denying freedom of movement to the Indian Peacekeepers and Military Observers deployed there. At the insistence of RUF all helicopter flights, including supply flights, were stopped to Kailahun and the garrison was road maintained by weekly ration convoy. The ration convoy was sent with the permission of the RUF commanders. The situation became precarious in Jul 2000 when even the ration was refused permission.

Op Khukri was a unique multinational operation launched in the UNAMSIL, involving India, Ghana, Britain and Nigeria. The aim of the operation was to break the two month long siege laid by armed cadres of the Revolutionary United Front (RUF) around two companies of 5/8 Gorkha Rifles (GR) Infantry Battalion Group at Kailahun by affecting a fighting break out and redeploying them with the main battalion at Daru. Its successful execution displayed the effectiveness of the United Nations and brought it kudos from every corner of the globe. The professionalism and dedication of the Indian Armed Forces was yet again on display for the world to see.

After the attack on Freetown, the ECOMOG gained control but were unable to defeat the RUF decisively. After negotiations, the Lome Accord was signed whereby all conflicting parties agreed to disarm, supervised by a United Nations (UN) Force. When requested by the UN, India as a responsible member of the UN, agreed to contribute troops, including an infantry battalion group, an engineer company and a medical unit to UNAMSIL.

While some tension always existed between the RUF and the Kenyans and Nigerians, events suddenly turned for the worse when on 1st May 2000 the RUF at Maken in the Kenyan Battalion (KenBatt) Area of Responsibility (AOR), attacked and overran UN forces. Due to a communication gap this information could not be passed to IndBatt in real time, as a result of which, on the morning of 2nd May when the Kailahun company commanders went to meet the local RUF commander about the planned disarmament rally, they were taken hostages. Certain Military Observers (MILOBS) present in Kailahun were also captured. While the capture of their commanders and MILOBS from 13 other countries made it difficult for the companies to take offensive actions against rebels, they manned their defences and steadfastly refused the RUF's demand for their surrender. Based on orders, the Battalion Second in Command (2IC) was despatched with a patrol from the Battalion Headquarters (HQs) at Daru, along with the RUF Cease Fire Monitoring Committee (CMC) member, to negotiate the release of the hostages at Kailahun. At Kuiva this patrol was stopped and surrounded by about 200 drugged rebels. As the battalion had experienced similar situations many times earlier during reconnaissance and initial deployment, the 2IC tried to calm the rebels down and began negotiations with their commander. The RUF commander requested the patrol not to go ahead since the situation was extremely volatile. He said that the RUF leadership at Makeni had informed all its cadres that the UN had attacked them. The patrol was detained and, while not being ill treated, and even being permitted delivery of food and movement of persons to Daru, was not allowed to leave as a whole. The hostage crisis at Kailahun was resolved 10 days later through intense pressure put on the RUF commanders by friendly civilians and the officers of INDBATT-1.

Necessity of Use of Force   The use of force became inevitable after all the diplomatic means had been exhausted to persuade the RUF to lift the seize on Kailahun Garrison. Troops had been confined to an area of about a square kilometre for almost 75 days and after the stoppage of helicopter flights

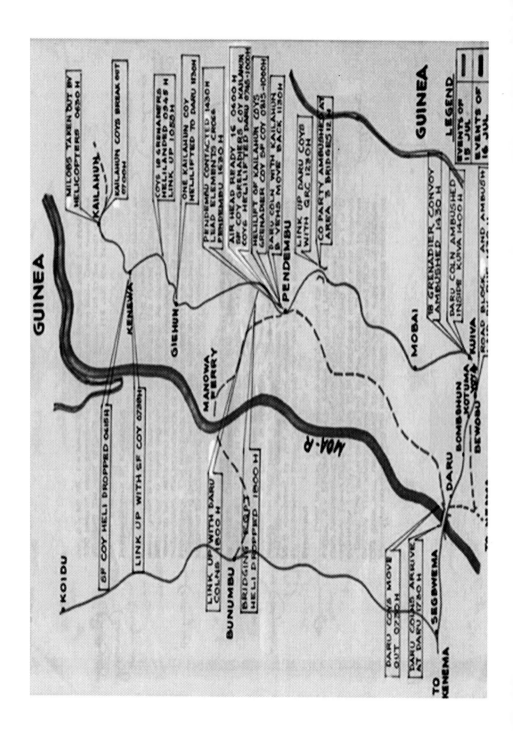

and road convoys by the RUF, the availability of rations with the garrison was fast depleting. The situation was further compounded by the intelligence reports that RUF was seriously planning to attack and disarm the troops to take them as hostages.

Force Levels   Although the operations were conceived and planned by Indian component of UNAMSIL and was to be executed only by the Indian forces, yet keeping in mind the international ramifications of the operations and to make it appear a UNAMSIL operation, token involvement of troops from Ghana and Nigeria was included in the execution phase of the operation. These troops were only utilized for filling small gaps in the defences of Daru which had come up due to Indian troops moving forward for the actual operations. One C-130 and two Chinook helicopters of Royal Air Force were utilized during the mobilization phase and subsequently to insert SF troops and extrication of Military Observers from Kailahun Garrison. Since the UN did not have adequate fixed wing and helicopter resources in the mission area and the speed and timings of the operation were considered to be of paramount importance, the British Govt was requested to assist in the initial phase of the operation by providing necessary air effort. This they had agreed on the condition that the British would be permitted to land at Kailahun to pick up Military Observers which included a British Officer before the first shot was fired towards Op Khukri. The composition of forces was as follows:

Formation	Mounting Base
Indian Army	
INDBATT-I (5/8 GR)	2 Companies(coys) at Kailahun/ 2 Coys at Daru.
INDBATT-II	(18 Grenadiers) Daru
Special Forces (2 Para)	Kenema
Quick Reaction Company	Daru
Artillery Composite Battery	Daru
Indian Air Force	
3 × Mi-35 attack H/C	Kenema
3 × Mi-8 Helicopter	Kenema
2 × Chetak Helicopter	Kenema
Royal Air Force	
3 × Chinook Helicopter	Kenema
1 × C-130 Aircraft	Kenema
Ghana Army	2 Coys at Kenema
Nigerian Army	2 Pl at Daru

Revolutionary United Front (RUF) Forces   The known dispositions of the RUF in the area of operations were as under:

No 1 Brigade HQ   Pendembu. In addition, the RUF maintained part of its GHQ at Kailahun and dominated road Daru-Kailahun by establishment of road blocks.

Battalion HQs:

Segbwema	Strike Battalion. Approximate strength 200-250 RUF cadres
Kuiva	Approximate strength 100-150 RUF
Mobai	Approximate strength 80-100 RUF cadres
Bunumbu Junction	Reinforcement battalion

Operational Plan

Phase 1   Mobilisation and move of combat elements from Freetown/ Hasting to Kenema/ Daru on 13-14th Jul 2000 by air/ road.

Phase 2   RAF Chinooks **to insert 80 personnel** of Special Forces short of Kailahun and extricate 11 MILOBS and 30 personnel along with 4 tons of stores from Kailahun at 0600h. Artillery (arty) engagements were to start from 0620h on 15 July at pre-designated targets. INBATT-2 was tasked to: Secure firm base for INDBATT-1 (Daru Column) earliest but not later than 0800 hrs on 15th July; Secure and hold Giehun by **helicopter assault** earlier but not later than 1200 hrs on 15th July. INBATT-1 less 2 Coys (Daru Column) was tasked to: Advance along axis Daru-Kailahun and link up with QRC column at area Three Bridges earliest but not later than 1400 hrs on; provide assistance (4 × BMPs) to INDBATT-2 in securing firm base. Kailahun based companies were to establish an **air head at Kailahun** to assist in air extrication of non-essential personnel, MILOBs and stores and prepare to break out by road link up SF team (80 personnel) earliest but not later than 0600 hrs. SF team were to secure general area South of Kailahun by **heliborne assault** earliest but not later than 0900 hrs. QRC column was to secure area Three Bridges by **helicopter assault** earliest but not later than 1200 hrs. Mech Infantry Coy was to secure Benduma earliest but not later than 1000 hrs. GHANBATT less two Coys was to advance from Kenema at 0630 hrs and secure Bendu Junction earliest but not later than 1200 hrs. One coy each NIBATT 3 & 4 to relieve company of INDBATT-1 and hold defended locality in Daru with effect from 1900 hrs on 14th Jul 2000. Two platoons ex-IND ENG COY-2 were to occupy a platoon defended locality in Daru with effect from 1900 hrs on 14th Jul 2000.

The attack Helicopters (Mi-25) were tasked to: carry out strikes at Kuiva, Mobai, Pendembu, Geihun, Manowa ferry, Kailahun and provide intimate fire support to Kailahun coy during breakout; maintain Continuous Over Watch (COW) over the column and provide fire support on call; provide air cover and suppressive fire for special heliborne operations and interdict RUF reinforcement columns. The Mi-8s were tasked to: airlift stores from Kailahun at first light plus 15 minutes and insert one coy of QRC at three locations in general area three bridges (South of Pendembu); to insert one coy of 18 Grenadiers North of Giehun and re-deploy SF team to general area North of Pendembu depending on progress of ops. Chetak helicopters were tasked to act as continuous airborne communication post, listen out on Motorola and VHF freq, co-ordinate comn between ground formations aircraft and base; and carry out recce and casevac on as required basis.

Phase 3   INDBATT-2 less a copy was to secure and hold Kotuma, Kuiva and Mobai earliest but not later than 1500 hrs on 15th. INDBATT-1 (Daru Column) with QRC was to secure and hold Pendembu and establish airhead by 1930 hrs on 15th. Kailahum column and SF team were to link up with coy of INDBATT-2 at Giehun by 1730 hrs. Mi-35 were to maintain Continuous Over Watch over the ground columns and provide firepower on call; also, to interdict RUF reinforcement columns and to be a threat in being. Mi-8s were to extricate foot column of Kailahun coy from airhead south of Kailahun and to provide logistic support on as required basis. Chetaks were to provide airborne communiction post and standby for recce or casevac.

Phase 4   Mech Inf Coy at Benduma was to be prepared to vacate the location at 0600 hrs on 16th, on order and act as force reserve. Link up of INDBATT-1 (Daru columns) and Kailahun columns at Pendembu was to take place by 1200 hrs on 16th. Air extrication of foot columns and non-essentials from the airhead establishment by INDBATT-1. Mi-35 and Chetak tasks same as in Phase 3.

Phase 5   INDBATT-1 columns were to advance and reach Daru by 1200 hrs on 17th. INDBATT-2 elements were to fall back to Daru after passage of INDBATT-1 columns. Mech Inf Coy at Bendurma was to return to Daru after vacation of Mobai on 17th in case not withdrawn from Benduma earlier. Mi-35 and Chetak tasks were to be same as in Phase 3 and 4. Mi-8 to provide logistic support on as required basis.

Execution: Air Tasks   Due to the involvement of IAF Contingent from the very beginning; adequate pre-warning was available and the aircraft servicing was suitably staggered to ensure availability of all aircraft during the entire period. To ensure degree of surprise, the mobilisation was carried out on D-1 and D-2. 21 pilots, 7 flight engineers and 4 flight gunners were involved. Spare crews were taken to cater for exigencies, fatigue and manning of operations room. Two technician gangs for Mi-35s were also planned. The strength of technical manpower was meticulously worked out to ensure availability of sufficient manpower round the clock.

A total of eight helicopters of IAF contingent participated in the operations; three Mi-35s, three Mi-8s and two Chetaks, They were flown to Kenema, by D minus 1 last light. Besides Indian helicopters, three CH-47s of British Forces were also positioned at Kenema by last light on D minus 1 to participate in operation. Airlift of the IAF detachment was affected by three shuttles of C-130 Hercules of British Forces airlifting 9-11 tons of load on each flight. It was envisaged that three sorties per aircraft per day would be required for three days. Based on the air effort required, ground equipment and tractor were air lifted to Kenema. Although 27 sorties were planned to be flown, armament was carried for only 18 sorties catering to the first two days of operation; with a stand-by plan to procure additional armament as and when the requirement cropped up. Armament calculations were done at the rate of 64 × 57 mm rockets, 500 × 12.7 mm front gun rounds and 64 flares per sorties. Kenema had bladder-refuelling facility; static ground refuelling however meant that the aircraft had to be taxied or towed to the refuelling point for each refuelling. Considering the expected intensity of ops and need to cut down turn round time, a bowser was requisitioned and was airlifted by a Mi-26 on D-1 to Kenema. The total projected requirement of fuel was 64 KL. One additional bowser was also placed at Daru.

RAF Aircraft   The operation commenced with RAF C-130 Hercules taking off from Kenema at 0527 hrs on d-day. The aircraft took off in pitch-dark night, and in rain, without any runway light, using NVGs and on board aids. Immediately after getting airborne, the C-130 established an aerial communication post utilising the state of the art avionics and secure communication contact with the Chinook aircrew waiting on ground and provided them with all the necessary inputs about weather and terrain. Equipped with the full picture about the weather and terrain, two Chinook helicopters took off at 0535 hrs with 80 SF troops of the Indian Company. All through their flight to the landing site short of Kailahun, the Chinooks were guided by the C-130 flying overhead besides their own precision navigation aids like INS, latest GPS and Doppler etc. They reached the landing site at 0600 hrs and slithered down all the troops safely at the designated landing site. Thereafter, the Chinooks continued and landed one after the other at 0610 hrs at Kailahun helipad in darkness and bad weather achieving total surprise. Within five minutes, both the Chinooks cleared Kailahun helipad after picking 11 MILOBS, 7 Indian troops and some load, and landed safely at Daru.

The first batch of helicopters sent to UNAMSIL, a set of four Mi-8s and Chetaks each are lined up for inspection at Palam New Delhi before being despatched to Sierra Leone. The Mi-35s were sent later when it was felt that more dedicated firepower was needed

IAF Helicopters   As planned, three Mi-8s and two Mi-25s got airborne from Kenema at 0620 hrs and headed for their respective destinations. However, immediately after getting airborne, they encountered rain with clouds touching tree tops and poor visibility conditions. When it became clear that the continuance of the flight was not possible due to extreme bad weather, they had to land back at Kenema. Weather conditions further deteriorated with the sun coming up and heavy down pour continued till 0800 hrs. At 0810 hrs, information was received from Wg Cdr AK Sinha based at Daru Joint Ops Room that there were signs of improvement in weather at Daru. At this point of time, although the weather was still bad at Kenema, a decision was taken to launch one helicopter for weather recce. This helicopter got air borne at 0820 hrs and finding the direct routing to Daru covered with dense fog and low cloud made a dog leg and proceeded to Daru following the river. The cloud base reported by this helicopter was 100 ft and visibility 500 mtrs or so in rain. Keeping in mind the urgent operational requirement of the mission, two more Mi-8 and two Mi-35 were launched after thorough briefing about the weather; and thus commenced the IAF helicopter operations towards Op Khukri. The main task of the Mi-8 helicopter was troop induction, re-deployment and de-induction besides logistics supply and Casevac.

Induction of Coy of 18 Grenadier at Giehun   Giehun village located on Kailahun-Daru road was a RUF Coy defended post. Its capture was critical to the advance of Kailahun column. One coy of 18 Grenadiers was to secure and hold Giehum by air borne assault. 73 personnel with 1000 Kg of ammunition were required to be helilanded. All three Mi-8s in one wave were utilized for the task. In the absence of any information about the likely landing sites, the helicopter was first required to locate a suitable site and then give the details to the following helicopters. On reaching the village area, the lead helicopter did not find any suitable site due to all trees and thick vegetation all around in the area of intended site. The only possible landing site was on the outskirts of the village, which was not considered suitable from the point of view of reaction from the rebels occupying the village. However, since there was no other choice, the helicopter Force Cdr agreed to land at this site which was about 30 × 30 m surrounded by 35 to 40 m tall trees all around. Immediately, the lead helicopter captain decided to land the troops and passed the GPS co-ordinates and LZ information to the other two helicopters. Subsequently, all of them carried out landing from OGE hover height of 35 to 45 m resorting to vertical descent. After all troops had been safely landed, there was exchange of fire between the rebels and our troops. However, there were no casualties and the mission was successful.

Induction of Indian QRC   Security of Three Bridges on the road between Pendembu and Mobai was critical to the progress of own motorized and mechanized columns. Hence, it was decided to induct a platoon of QRC at each of these bridges. In the absence of any suitable landing site, the troops were landed on the road itself next to the bridges. The landing site at the bridges were similar to the one at Giehun with just 5 metre rotor clearance from 25 to 35 metre tall trees which necessitated a very precise descent from OGE hover and landing on the road barely wide enough to accommodate both the main wheels. The lead helicopter after locating the Landing Zone (LZ) passed the GPS co-ordinates and other LZ information to the other two helicopters following behind. 141 troops and 3450 kg of ammunition and other supplies was safely landed at these locations successfully.

Re-deployment and Extrication   After the QRC troops had been inducted at the Three Bridge area, the rebels reacted and an attack on Bridge Three became imminent. Hence, it was decided to redeploy two QRC platoon including mortar platoon at Bridge Three. Two Mi-8s were utilized and this was accomplished before rebels could overrun the location. As planned, after the successful link up of Kailahun column with 18 Grenadiers at Giehun, all the foot columns of Kailahun Coy along with three casualties were airlifted from Giehun by utilizing three sorties by Mi-8s. The area was secured by the ground troops and the extrication of troops from Giehun was uneventful.

Logistic Re-supply and Bridging Stores   An urgent message was received from ground troops that they were in need of water. Immediately one helicopter was launched and water had to be dropped from a height of about 25 metres since there was no landing site at the location of the moving troops eight Km short of Pendembu. Subsequently, the helicopter carried out a recce of the road between the locations of own troops and Pendembu and spotted digging of the road by the rebels for ambushing our troops 5 km short of Pendembu. This information was immediately passed to our troops who moved to either sides of the ambush site thus averting a serious disaster. The digging of the road by rebels five km short of Pendembu became a major obstruction for our Kailahun troops approaching Pendembu, and there was a critical need of bridging. Immediately, bridging equipment was arranged and one Mi-8 helicopter was launched to carry it to the site. Finding no place to land, the equipment was free dropped from OGE hover height of 25 to 30 metre, which enabled quick movement of troops to Pendembu just in time to prevent attack by the rebels chasing the column.

Extrication of Troops from Pendembu   By 2000 hrs on D-Day, all the troops moving from Kailahun and Daru had linked up at Pendembu and spent the night there, since night movement of troops towards Daru was considered dangerous due to non-availability of air cover by night. In order to ensure expeditious movement from Pendembu to Daru on D+1, it was planned to extricate all the foot bound troops by helicopters leaving behind only motorized troops to move by road. A total of 313 troops including two casualties were airlifted in 13 sorties utilizing 3 × Mi-8s. During the period of extrication of troops some rebel movement was reported north west of Pendembu which was tackled by a Mi-35 flying in the vicinity. The quick extrication of foot-borne columns from Pendembu by a Mi-8 facilitated faster movement of motorized columns to Daru; under the ever-present air umbrella of the Mi-35.

Attack Helicopters (Mi-35)   In case of well defined, pre designated targets, Lay Down Attacks in the First Run (LDA-FRA) were carried out employing 57mm rockets and 12.7mm front gun. In some cases, where the presence of shoulder fired heat seeking missiles was suspected, IR flares were also used. The details of tasks carried out by the attack helicopters is as given below:

- **Pre-emptive Strikes along the Axis of Advance** Two Mi-35s got airborne at 0830 hrs to carry out strikes on RUF strongholds along the axis of advance. One helicopter engaged pre-designated targets in Bewobu, Yoya, Kuiva, Mobai and Pendembu from south to north. While the other attacked Manowa ferry and Giehun before establishing communication and visual contract with the ground column. This helicopter then engaged certain pre-selected targets in Kailahun after confirming the location of own troops with SF Coy Cdr. These strikes involved lay down attacks in the first run (LDA-FRA) employing 57 mm rockets, 12.7 mm Front Gun and IR flares as required. The helicopters were fired upon with small arms from Kailahun, Manowa ferry and Pendembu. However, there was no damage to any of them.

- **Suppressive Fire at Mi-8 LZs**   The pre-emptive strikes terminated into speculative suppressive fire at Giehun and Pendembu DZs, just prior to troop insertion by the Mi-8s. Subsequently, a random pattern for air patrol by Mi-25s was resorted to. The suppressive fire and subsequent presence of Mi-25s in the vicinity of DZs ensured that the drops took place without any interference from the rebels.

- **Interdiction of Enemy Reinforcements**   During the planning stages of the operation, potential RUF strongholds that could have interfered with the advance were identified. These places like Buedu, Segbwema, Bunumbu Junction and Koidu were connected by tracks heading towards the axis of advance. In order to preclude possible build up of reinforcement from these places, timely attacks were carried out to soften them up.

- **Armed Recce/ Convoy Protection**   This was one of the major tasks of the Mi-35 helicopters. The intention was to act as rearguard and to prevent any interference from the flanks or from ahead. Considering the fact that our troops were bound to the only road link from Kailahun to Daru, it was feared that the rebels would create road blocks to impede the advance. In addition, rebels started attacking from the rear. However, with the arrival of attack helicopters, the convoy did not have to put in any effort in guarding its rear and only concentrated on its advance towards Daru. The coordination with which the Mi-35s provided convoy protection was a result of effective communication between the ground forces and the pilots. During the conduct of all the sixteen missions, spread over two days, positive two-way contact was maintained at all times on VHF/ Motorola sets provided by UN.

Light Utility Helicopters (Chetak)   Two Chetak helicopters positioned at Kenema took part in Op Khukri. Both these helicopters were utilised to provide continuous airborne communication post over the mission area and established a relay centre for low flying helicopters and ground troops. Casevac and communication tasks were also assigned to them. The round-the-clock airborne communication post proved to be very effective in maintaining touch with the Joint Ops Centre, all the columns of ground forces and the Mi-8 and Mi-35 helicopters. This facility enabled minute to minute monitoring and control of the air ground operations besides ensuring immediate air support to the ground forces on call. Keeping in mind the threat from rebel forces and the cone of fire of own artillery and mortars, the loiter point and altitude of the Chetak helicopters was modified. They were utilised for undertaking communication sorties for the Force Commander who landed at Giehun and Kuiva on D-Day amidst the battling troops and at Pendembu on D+1. A Chetak evacuated a casualty from Giehun on D-Day and two casualties on D+1 Day from Kuiva. Casualties from Kuiva hit by RPGs were evacuated amidst a heavy battle zone. Within five minutes of the occurrence of casualty, the helicopter landed on a very small clearing under very poor visibility conditions created by the bellowing smoke as a result of firing of heavy weapons. The casualties were on the operation table at Daru within fifteen minutes. Further, these helicopters proved very useful in the scout and airborne FAC role, in addition to providing early warning to ground troops about approaching threats from rebels and roadblocks. Motorola as well as VHF sets were utilised to maintain uninterrupted communication with helicopters, ground troops and Joint Ops Centre.

Details of Flying Efforts: MI-35

Sl No	Day	Sorties/Hours	Armament expended		
			12.7mm	RPs	Flares
1	14 Jul 2000	03/03:00	-	-	-
2	15 Jul 2000	08/15:00	4472	509	45
3	16 Jul 2000	08/14:15	3434	409	139
4	17 Jul 2000	03/03:30	-	-	-
	Total	22/35:45	7906	918	184

For 75 days, the rebels held 222 peacekeepers hostages and the situation had become precarious. The UN Force Cdr took the military option. The IAF contingent was tasked to execute the extraction with a host of restraints placed on them. This mission was brilliantly accomplished along with UNAMSIL forces which consisted largely of Indian Peacekeepers. During the operation, the IAF Contingent flew a total of 98 sorties utilising 66:05 hours. A total of eight helicopters were utilised for this mission.

Appreciations

Lt Gen NC Vij PVSM UYSM AVSM, DGMO (10th Jun 2000). "Great Meeting you all. Well done. You have done the country proud and earned tremendous goodwill. Wish you, the very best of luck and adventure and glory in days to come".

Maj Gen VK Jetly UYSM, Force Commander (10th Jun 2000). "You all have, in a short span of time worked wonders and earned a niche for yourself internationally. As the Force Commander, I am extremely proud of having such an outstanding professional outfit under my Command. Keep up the good work and God bless".

Dr. C Halle, DPKO, UN HQ, New York (16th Jun 2000). "I am deeply impressed by the ability of your staff to see solutions and not problems in their task in the true spirit of the United Nations".

Reasons for Success

- **Synergy of all Assets**   Op KHUKRI was a classical example of synergy of effort. The optimum utilisation of all resources, joint planning (Indian Army, IAF, UNAMSIL forces and the British Forces) and execution resulted in a synergy that multiplied the effectiveness of the assets deployed. All types of helicopters played their key parts in the operation.
- **Simultaneity of Operations**   Commencing operations simultaneously from Kenema, Daru and Kailahun, and heli-landing troops at three places enroute caused utter confusion in the RUF.
- **Real Time HUMINT and SIGINT**   During the planning of rescue mission from Kuiva, Pendembu and Kailahun, non availability of air/satellite imagery was a big handicap. An NCO of 5/8 GR made nine trips with the ration convoys to covertly note RUF strength, deployments and obtain GPS fixes of RUF targets. This intelligence was later used by the artillery and attack helicopters to engage targets. The Battalion radio monitoring cell did an outstanding job in monitoring RUF communications throughout the three months to build up a clear picture of the RUF activities. Real time monitoring of RUF communications during operations gave a picture of RUF actions regarding move of reinforcements and preparation of ambush/road block sites. These were passed on to the columns and to attack helicopters for verification and engagement. The painstaking study of RUF tactics, organisation, and personalities of leaders, and updating of activities helped to anticipate reactions correctly.
- **Deception and Security**   RUF was successfully kept in the dark about the build up. The RUF was made to believe that the battalion was being replaced, and hence additional troops were coming as

relief. No operational messages were passed on radio and only satphones were used. The unit had made own codes and nick names for places and personalities. All conversation was in Hindi and Gurkhali; and for further secrecy, plans were passed over satphone to Kailahun in colloquial Malayalam. The plan had an inherent capacity to be changed as per the progress and situation. The presence of the Force Commander at Daru helped further, as decisions were taken on the spot.

- **Attack Helicopter Support**   The attack helicopter was one of the most potent assets and proved very effective in breaking ambushes as well as denying free movement of rebels on the road by day. Combat Air Patrol (CAP) provided to the returning road column on 16th July 2000 was instrumental in its safe return.

The success of Op Khukri was felt not merely in its tactical terms. It gave the RUF its worst defeat in recent history and at the same time gave a tremendous boost to the UNAMSIL forces in particular and to the UN as a whole. Detractors of the UN were silenced and the potential of this noble institution was once again displayed. The greatest reward for Indian Peacekeepers was the reception given to them by the people of Daru as they came triumphantly back from battle. It was the welcome of a long suffering and desperate people who understood that there were others in this world who would shed their blood for their just cause. Perhaps, that is why they helped build the Khukri Memorial in Daru barracks overlooking the Moa River.

Other Operations in UNAMSIL

Rescue Operation-Makeni   On 7th May 2000, the IAF contingent was tasked to evacuate three Kenyan battle casualties and 11 UN Military Observers from the besieged garrison of Makeni, a major town in the north east of Free Town at a distance of about 150 Kms. Immediately after landing at Makeni helipad, the helicopter piloted by Sqn Ldr TA Dayasagar and Flt Lt MK Yadav, encountered heavy firing from the RUF rebel position around the helipad. The crew, undeterred by the firing, offloaded the supplies at the helipad and the casualties and the terror stricken Military Observers into the helicopter and took off with the rebels still firing at the helicopter. As soon as the helicopter was airborne, it developed very heavy vibrations as a result of damage and almost became uncontrollable. It was flying over populated and forested areas with no force landing field in sight. The crippled helicopter was flown almost 10 Km before being put down on the first available clear patch. The emergency "May Day Call" was picked by another helicopter on similar mission piloted by Wg Cdr RK Negi, who immediately headed for the force-landing site. It soon landed next to the crippled helicopter and quickly shifted all the casualties and military observers. During this period, the rebels were seen approaching the helicopters from all directions. Under these circumstances with imminent threat of a rebel attack, the crew had to take the decision to abandon the crippled helicopter. Subsequently, the crew, casualties and the military observers were rescued safely to Hastings. This daring and well coordinated rescue operation earned wide appreciation from all including from UN HQs, New York.

Emergency Supply of Ration and Ammunition   On 10th May 2000, an emergency call was received to supply rations and ammunition to Kenyan Battalion, which was moving from Makeni to Kabala. In the absence of any landing site in the area, the Force HQ was obviously in a predicament to deliver the much needed supplies to the Kenyans. A Mi-8 landed the supplies safely into the hands of Kenyans in the rebel dominated area. It will be relevant to mention here that the aircrew were given only approximate coordinates of the rendezvous with the Kenyan troops who were on the move and it was quite difficult to locate them in the thickly wooded terrain. Notwithstanding the numerous odds, the contingent aircrew lived up to their high professional standards and located the place and earned the admiration of the Kenyans.

Modification for Armament Role   During the crisis situation created by RUF rebel attacks on UNAMSIL positions, both Mi-8s and Chetaks were modified to provide cover effectively from the air. All these modifications, innovations and challenging high risk missions were undertaken at a time when UNAMSIL forces were under extreme pressure from strong rebel attacks. The fire support provided by helicopters created panic among the rebels and at the same time boosted the morale of UN forces, finally blunting the rebel offensive. The contingent aircrew also took up another challenging task of providing aerial reconnaissance and acting as mobile command posts in sensitive mission areas.

AIR POWER

United Nations Peacekeeping Missions

Indian Aviation Contingent - (Goma)

MONUC was the third major contribution of the IAF under UN flag, after Somalia and Sierra Leone. As the situation in North-East Congo turned grave with repeated massacres and killing of innocent civilians, the international community decided to strengthen the military presence. India contributed armed helicopters and utility helicopters in the Congolese provinces of North Kivu and Ituri. The IAF unit in Goma/Bunia was called the Indian Aviation Contingent (IAC-I) and its major lodger units included IAF Squadron 2003 (Vipers: Mi-25s) and IAF Squadron 2004 (Equatorial Eagles: Mi-17s). Vipers were equipped with four Mi-25 aircraft while Equatorial Eagles had five Mi-17 helicopters along with aircrew and ground support personnel. These invaluable assets increased MONUC's sphere of influence in the Eastern DRC; and UN forces were able to reach areas that had so far been outside its sphere of influence.

Operations   IAC-I had been in the forefront of all the offensive support and humanitarian tasks of MONUC since induction. Some of the major operations in which the IAC-I took part were Op North Nationalism, Op North Necktie, Op North Nuclide and Sake Ops, Op Kimia-2, Op Rudia-2 (Bas-uele & Haut-uele), Op Amani Leo, Op Iron stone, Ituri engraver and Op Western thrust. In these operations casevac, mounting of HOBs, Helicopter Landed Aerial Domination Patrol Missions (HLADP Missions) and all kinds of support operations were executed. These operations ensured time and again that the vital towns of eastern DRC did not fall to rebel forces. The number of complex operations jointly conducted along with the Indian Army and other armies to handle battle-hardened rebels exceeded more than 150 in eight years. The following are some examples:

- *OP Fataki:* On 4th Sep 2003, IAC-1 attack helicopters along with BANAIR Mi-17s took part in a joint mission to induct ground recce forces to assess the situation in the village of Fataki. In the mission, one of the attack helicopters (AH) inserted a helipad securing team along with Chief of Staff of the Ituri Bde. This was an exception to the rule, as the BANAIR elements refused to land on the grounds of unfamiliar area and unsecured helipad.

- *Op Shomiya:* On receipt of the intelligence report that FNI elements were planning an attack on Shomiya village at 0500 hrs on 31st Oct 2003, one full platoon of BanBatt was airlifted from Bogoro village located SE of Bunia town by two Mi-17s to the village along with Mi-25 as escort. The proactive role of Mi-25s not only thwarted attempted attack by militia but also prompted surrender of arms.

- *Op Shahbunda:* This mission was conducted in three phases, wherein phase I & II were conducted in Sep 2003, which culminated in signing of peace agreement between ANC and Mai-Mai on 1st Oct 2003. The agreement was aimed to control the illegal mining activity in the mineral rich area of Shabunda and Lulingu. However, despite the agreement, it was found that the agreement was not being followed in spirit and illegal mining persisted. Accordingly, IAC-1 acted as force multiplier in the joint operations to facilitate lasting peace in the region.

- *Op Bogoro:* Fire support (57mm rockets) was provided by Mi-25s on 2nd Nov 2003 to the BANBAT located at Bogoro situated 15 km south of Bunia. The UN troops were reported under fire from armed militia. On 4th Dec 2003, a Mi-25 fired 29 rounds on request from the UN ground forces which had surrounded the militia, who were returning fire and were not getting subdued at village Nyakunde (40 Km south of Bunia). The impact of Mi-25 front gun firing made the militia run for cover, hence giving immense advantage to the ground forces.

- *Op Kasango/Op Shabunda:* On 11th Sep 2003, two Mi-17s carried out recce and insertion/extraction of 169 troops, MILOBs and UN negotiators at Kasango and Kampene. On 23rd Sep 2003, two Mi-17s carried out insertion of 36 MILOBS and troops at a site which was suitable for only single helicopter operation. The Mi-17s landed at an interval of two minutes and disembarked the troops who secured the helipad. In this operation, one Mi-25 provided cover, while two Mi-17s disembarked the 55 personnel for deployment at Shabunda. On 25th Sep 2003, 35 personnel were transported from Kindu to Kama for a night to hold talks with local leaders. Mi-25s sanitised the entire area of operation.

- *Underslung Ops:* For the first time in MONUC, an underslung sortie with a generator weighing 1230 Kg from Goma to Nyabiyando was deployed during Sep 2003 in support of the South African Company.

- *Op Kishanga/Lubutu:* On 25 Sep 2003, a negotiating team was inserted to Kishanga alongwith Task Force 1 troops. This enabled negotiations with warring factions of RCD (G). On 22nd Nov, a team of 10 personnel of TF-1 troops along with DDRRR personnel were deployed at Lubutu and Mubi by two Mi-17s to carry out the peace process with rebels.

- *Op Mohanga/Fizi:* Timely insertion of a negotiating team by Mi-17 with an armed escort (Mi-25) at Mohanga on 18th & 25th Nov 2003 enabled negotiations and surrender of arms by rebels. On 28th, three Mi-17s were pressed into action to induct troops and DDRRR teams to Fizi. The mission involved operation beyond the normal radius of action entailing provisioning additional fuel in internal tanks and thorough planning for safe and efficient operations. During the operation, one Mi-17 was stationed at Uvira (intermediate place) for additional fuel.

- *Op Kilungutwe:* On 27th Jan 2004, Mi-17s of IAC-1 undertook airlift of UN troops and equipment to Kilungutwe with a Mi-25 as escort. However, with the Mi-25 getting unserviceable, a Mi-17 made the operation a big success after taking on the armed role (after obtaining the prior sanction of UN HQ New York), despite the armed rebels sitting in the firing positions on the dominating hills. The entire mission led by three Mi-17s in a determined show of force (with one helicopter armed with rockets) was duly applauded and the BBC termed it as one of the biggest humanitarian aid missions undertaken in the area in recent times.

- *Op Eka Barrier:* On 25th Jan 2004, front gun and rocket firing by Mi-25 was sought by the UN ground forces which had come under fire from the renegade elements. This had a dramatic result leading to immediate retreat by militia. On 12th Feb, a MILOB team returning from a verification mission was ambushed near village Katoto. One Kenyan Major was killed in the ambush and rest of the team had to abandon the body as well as the vehicles. On receiving the input, Mi-25 was pressed into action, which continued to thwart various attempts by the militia to capture the vehicles as well as the body of the MILOB, till such time the UN ground troops took charge.

- *Op Niyamamba:* On 24th Feb 2004, UN troops sought for a fire support from Mi-25, when they were ambushed at village Niyamamba. 28 rockets were fired and the militia camp was partially destroyed by a Mi-25 and all assistance was provided to the ground forces till it returned safely. Again on 18 Mar 04, Ituri Bde planned a proactive attack on militia group in general area of Niamamba. Ground forces of UruBatt & BanBatt and air elements of BANAIR & IAC participated in the operation, wherein Mi-25 fired 42 rockets in support. Three militia camps were totally destroyed, four militias killed, one injured and 24 were apprehended.

- *Op Drodo:* On 27th Mar 2004, Mi-25 fired 93 rockets in support of UN ground forces, which had launched offensive in Drodo village to locate the Fataki Chief (who had been abducted by Bosco's group and kept in custody in the hospital of Drodo). The extensive fire power, not only made the militia run for cover but it virtually razed to ground two militia camps and Bosco's house. On reported killing of civilians by militia in Mubanga town, a Mi-25 was tasked along with NepBatt troops in APCs to control the situation. Active participation and proactive action by the Mi-25 not only identified the militia (more than 60-65 in number) hiding in the woods about 2 km from village, but also fired 14 × 57mm rockets in two salvos to blast out the militia.

- *Op Lusamambo:* On 11th Feb 2004, overnight preparations for an early morning reinforcements and subsequent de-induction by Mi-17s was necessitated from a small village (Lusamambo) located 55 miles NW of Goma, as there was an imminent threat to UN ground troops. There were lot of excesses reported against the locals. The quick reaction and response from IAC team won accolades from various quarters, wherein all preparations for the mission commenced at 0400 hrs on 11th Feb 04. Again on 18th Mar, due to the reported planned attempt by militia to attack the Presidential Palace at Kinshasa, the deployed troops (in Lusamambo) with entire equipment was promptly extricated with deployment of all available five Mi-17s.

- *Op Mwenga:* Mwenga is an area SW of Bukavu and has been a hub centre of various militia groups, who were obstructing MONUC's peace initiatives in the area and were also involved in several cases of human right violations. The MILOB team, which was sent to negotiate on 16th Feb 2004 with armed Mai-Mai militia and other warring groups, met hostile resistance. The situation had become tense, wherein one of the groups wanted to take the MILOB team hostage. This was retrieved by prompt retrieval of the team by Mi-17s accompanied with a Mi-25. The success was applauded by Chief of Staff, Bukavu and it highlighted swift and timely action by Mi-17 pilots, who reached the target 30 minutes before the expected time.

- *Op Virunga:* This operation was carried out from 20th to 24th Apr 2004, to not only put to rest the infighting among the belligerent groups in the Virunga National Park, but also identify the presence of Rwandan Troops (Army). The mission was a great success as the presence of approx 4,000 Rwandan soldiers inside the Congolese territory was confirmed by the UN. This led to outcry by the UN and more pressure was mounted on the Rwandan Govt to immediately withdraw from DRC region.

- *Op Kwandroma:* On 13th Sep 2004, two Mi-25s were tasked to provide fire support to Ituri Bde troops in cordon and search operations against FNI troops in Kwandroma and adjoining villages. On 28th, a Mi-25 was tasked to bail out the PAKBAT troops who had come under heavy fire from militia close to Nizi. 25 militias were securely hiding about 250-300m from PAKBAT and were resorting to firing. However, on show of force and offensive manoeuvres by a Mi-25, the group dispersed in the jungle. Again on 21st Nov, a Mi-25 was launched to provide fire power to support PAKBATT troops, who were under attack by militia at their HQ at Nizi. Since, the militia had securely hid in houses in the centre of town, the AH fired two warning rockets, which deterred the militia from firing.

- *Op Ex-Mahagi ALG:* Activation of Mahagi ALG was made during Sep 2004, to significantly increase the radius of action and time over target of Mi-25s basically to provide adequate coverage to the Aru sector, which is 160 Km from Bunia. Non-availability of intermediary refuelling station in the past restricted optimum utilisation of Mi-25s in the sector.

- *Op at Ndrelle:* On 5th Dec 2004, two Mi-25s were tasked to provide fire support to Ituri Bde troops carrying out CASO at Ndrelle. It initially fired two rockets at the military camp as warning shots. But when the militants refused to compromise, Mi-25s were tasked to neutralize the camp. Accurate and effective firing apart from cover helped ground forces to carry out mopping up operations.

- *Op in Ituri Sector:* Extensive operations in the Ituri sector were undertaken to quell the militia during Jun 2005. Major operations were carried out at Lugo, Balu, Boga and Medu areas of Bunia. On 8th July, extensive support of Mi-25 was sought by MECHEM (De-mining team) in Kagaba – Aveba area (south of Bunia), when it came under heavy firing from militia. Again, on 16th, when militia struck the BANBATT company during its withdrawal from Aveba, attack helicopter assistance was sought to disperse the militia group.

- *Op Kivu:* On reports of clashes between forces of 10th Mil Region and 8th Mil Region to gain control over mineral rich areas on border between North and South Kivu apart from probable advance of forces of Gen Kunda from Walikale (via Hombo), Urubatt and RSABatt were tasked to conduct area domination missions from 24th– 29th Sep 2004. Towards this task, Mi-17s carried out insertion as well as extraction sorties in the area from Walikale, Hombo, Minova, Masisi, Numbi, Tsunguti and Nyabiondo.

- *Op Mutwanga:* Mi-17s were tasked during Sep 2004 to carry out insertion of troops and DDRRR personnel for recce and verification tasks in a small town east of Beni. The salient feature of this area was the helipad, which was located in the foothills and had a very restricted approach funnel due to presence of tall trees.

- *Op at Kamanyola:* On 10th Oct 2004, an urgent requirement to insert troops from Goma to Kamanyola (a small township located south of Bukavu at the confluence of Burundian, DRC and Rwandan border) was sought to achieve tactical advantage for the UN forces. Active and proactive response of IAC helicopters was duly appreciated by MONUC authorities.

- *Op at Walikale:* During Oct 2004, extensive sorties (64 in number) were flown in Walikale region (a village located in the mineral rich area west of Goma) to influence the militia by show of strength as there were frequent reports of infighting among various groups of militia to assert their control. Insertion and extraction of UN troops from Goma as well as Bukavu were undertaken.

- *Op at Bukavu:* On 6th Oct 2004, on receipt of the message that UN HQ building in Bukavu was under attack by local populace, a Mi-25 was immediately pressed into action to stop any such attempt. A large crowd was observed just outside the HQ of 10th MR, which subsequently disbursed on noticing the Mi-25 in an aggressive mood. On 18th Oct 2004, Mi-25s played a vital role in averting a major crisis on receipt of message that the Banyamulenge refugees returning from Burundi under UN escorts were under threat of attack by local populace at Uvira. Mi-25s not only escorted the convoy for the entire route, but also provided much sought after coverage and fire support to the camp till such time reinforcements arrived.

- *Op at Walungu:* The area in and around Walungu (a village located south east of Bukavu) under the control of FDLR militia was reporting regular looting and killing incidents. FARDC troops were unable to contain the situation. IAF carried out 104 sorties towards insertion of UN forces and equipment during Nov 2004 to control the situation. Mi-25s were tasked for CASO (cordon & search operations). Op Eagle was carried out during Aug 2005, wherein the main thrust was towards insertion/extraction of troops in Lusamambo. Op Falcon Sweep was carried out in South Kivu, wherein Mi-25s provided fire support to ground troops and Mi 17s at Miranda and Betara.

- *Op Ituri Energise/Eden:* Ituri Bde launched this from 20th-28th Dec 2005 along Fataki–Nioka–Mahagi axis against an armed group headed by Peter Karim Udaga, a former leader of the predominantly ethnic Lendu militia of FNI to clear the axis of militia activities. Mi-25s supported the operation in area domination, armed recce and fire support. Joint Op of the Ituri Bde along with FARDC was conducted in Boga-Aveba area to flush out militias and to stop infiltration of ADF / NALU from Beni sector after Op North Night Finale. This was terminated on 29th Dec 2005.

- *Op North Nexus:* This major operation was launched during 31st Oct 2005 by North Kivu Bde in Virunga National Park, the area adjacent to Rwanda, to make it a weapon free zone and to disarm the FDLR / FOCA and Mayi-Mayi militia in the area. The Indian battalions were tasked to compel the militia to surrender, lay down arms as well as forcing / coaxing them to join the peace process (DDR / DDRRR to join main stream of armed forces by mixing of troops in integrated Brigades of FARDC). A four helicopter (Mi-17) SHBO was undertaken on 31st Oct 2005 towards this operation. Mi-25s played a pivotal role in this operation. 304 militias surrendered and 152 arms were confiscated.

- *Op Aru-Aba/Op Ituri Eagle:* In the Ituri sector, a Ugandan rebel group known as LRA (Lord's Resistance Army), a rebel group operating in North Uganda and South Sudan for past 19 years, responsible for innumerable atrocities and human right violations, moved into Aba area

from southern Sudan during second half of Sep 2005, when chased by the Ugandan Army. Mi-25s played an important role in support of FARDC, Nep Batt and BANAIR in taking them on.

- *Op Eastern Eagle:* This was conducted in SE area of Lake Albert in general area Similiki from Bunia between 7th and 25th Oct 2005. It was aimed to apprehend the arms and ammunition being ferried by boats across Lake Albert from Uganda to DRC. Guatemalan SF and BanBatt troops from Bunia / Kasenyi were deployed to Niamavi by Mi-17s with cover from Mi-35s.

- *Op Ituri Enthusiast:* A joint operation consisting of Ituri Bde, FARDC and IAC-1 was launched in Nov 2005 to flush out militia from Similiki area (south of Bunia).

- *Op North Night Finale:* This offensive Op was launched from 23rd to 27th Dec 2005 to dislodge the cadres of ADF NALU (a Uganda based international organisation), which was posing a threat to stability and security in the NE parts of DRC. *Ind Batt –II, lost one JCO (Nb Sub Ram Kripal Singh) during the operation against the ADF / NALU cadres on 25th Dec.* Extensive air support missions were launched during the operation. During the operation, hover jump of 36 INDBATT-II troops was carried out at Beni ALG as show of force.

- *Op North Neon & Op North Nutshell:* This Op was carried out during Jan 2006 by regular insertion/extrication at various helipads / LZs in Rutshuru and Kanyabayonga sector due to road blockage near Sake. Op Paper Tiger was launched from 8th to 23rd Jan 2006 in Aba sector to clear the area of LRA militia, who had taken refuge in DR Congo. SF forces from Guatemala helicopters from IAC-1, IAC-II and BanAir participated.

- *Op Ituri Encourager:* A joint operation consisting of forces from Ituri Bde, FARDC and IAC was launched in Jan 2006 to flush out militia groups from Aveba and adjoining areas SW of Bunia. On 18th Jan 2006, during a fire support sortie to PakBatt, IAC helicopter (Z-3128) came under fire from militia and sustained two hits. One bullet passed through Captains canopy and other one punctured a fuel and hydraulic line on port side. The crew fired 27 rockets on the militia in three passes.

- *Op Ituri Engraver:* As a sequel to Op Ituri Encourager, Op Ituri Engraver was launched on 22nd Feb with an aim to flush out militia from their hideouts in area SW of Bunia. A three pronged attack was planned by Ituri Bde on militia stronghold at Tchei (SW of Bunia) and attack helicopters were to provide fire support. Ground forces succeeded in gaining mileage over the militia during Feb 2006, however, due to a revolt of FARDC Commando Battalion, MONUC suffered a sharp reversal of fortunes on 1st March. The Mi-25 detachment rose to the occasion and established a record on 2nd by providing a continuous cover / watch over the area from sunrise to sunset. Nine sorties for continuous 14 hours and 30 minutes over-watch and firing of 118 rockets in a day was achieved.

- *Op Ituri Element:* This was launched to give fire support to NepBatt between Nioka-Fataki-Kwandroma-Dera sector, covering areas of Dera, Lana & Guju (45 Km SW of Kwandroma), where Peter Karims FNI militia group was active. Sgt Adhikari Gyan Bahadur of Nepbatt was killed and three other injured during engagement with FNI militia in the Nioka – Dhera – Lana area of Ituri sector on 28th May 2006.

- *Op Ituri Idyll/Impulse:* During this Op, Mi-25s were pressed into action, when militia suddenly attacked the BanBatt troops on their patrol to flush out militia hiding in the hills close to Tchomia. The Mi-25 had to fire rockets in support of Banbatt. On reports of militia trying to get arms and ammunition supplies through Bohuma area to accelerate its activities against MONUC and FARDC, Pakbatt was tasked to carry out special CASO operations in the areas of Kasenyi, Tchomia, Katoto and Loga to neutralise the situation. Mi-25s provided air support to NepBatt during May 06, which was tasked to carry out CASO in the general areas of Nioka and Dera to neutralise militia (FNI militia group of Peter Karim) activities to restore peace in the area.

- *Op Ituri Ember:* The militia stronghold in general area of Jookafwe and Bule mountains (areas east and NE of Bunia adjacent to NW coast of Lake Albert), which gave them free access to shores of Lake Albert opening corridor for re-supplies, was to be secured by UN forces. The operation was launched on 9th May 2006. Op Ituri Explorer was launched on 20th May, when the desired results of Op Ituri Engraver were not achievable due to reverses faced by UN forces during the month of March. Ituri Bde launched an offensive in conjunction with FARDC at Tchey area with the view to destroy the military positions in the AOR. IAC dett played a vital role to plug the communication loopholes. This was extended up to 6th July.

- *Op Ituri Element–III:* A CASO by NepBatt along with FARDC and MorBatt was launched on 28th May to control militia activity in the Nioka–Kwandruma area, where increased military activity was reported. The NepBatt foot patrol came under heavy fire from militia, seven Km north of Fataki and seven NepBatt soldiers were taken as hostages. During Jul 06, in one of the fire fights with militia, BABBATT lost one of their soldiers and three were injured. Of the seven hostages, two Nepbatt soldiers were released by militia, and another two during Jun 06. The rest five were released after prolonged persuasion by MONUC on 8th Jul 2006.

- *Op Kumbokabo:* On 7th May 2010, this mission was launched by Mi-25s in support of the ground forces, which had come under unprovoked heavy fire from militia groups. It all started when a BANBATT convoy was fired upon while moving from Bunia to Marabo on Bunia–Beni road near Kombokabo (20 Km SW of Bunia). The Mi-25 operation in three phases on the same day at 0905 hrs, 1125 hrs and finally at 1430 hrs helped out the BanBatt and NepBatt troops. 88 rockets apart from front gun bursts immensely helped the UN ground forces. Apart from three bullet injuries to a Warrant Officer of NepBatt, Ten militias were killed and extensive damage to the militia camps was reported to the MONUC and UN HQ.

Indian Aviation Contingent-II (Bukavu)

The Indian Aviation Contingent-II (IAC-II) was inducted in Feb 2005; with its fleet of six night capable Mi-17 helicopters at Kavumu Airport in Bukavu

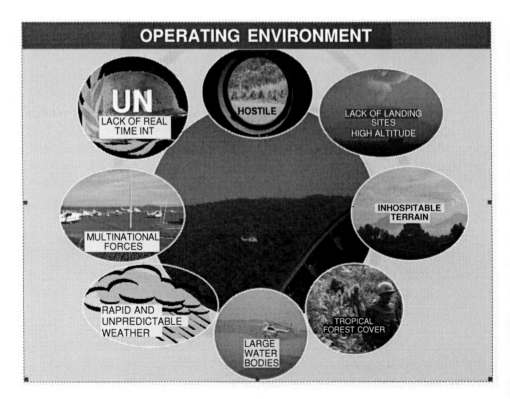

and four night upgraded Mi-35 helicopters at Goma, alongside IAC-I. Though it was not the first time that Indian Air Force was fielding a contingent abroad, it was certainly the biggest, with 10 helicopters, 36 vehicles and initial strength of 285 personnel. The MoU between the UN and Govt of India this time necessitated a fully self-sustained contingent in terms of electricity, water supply, communications, transport and any other possible requirements. This was indeed an immense task and the personnel of IAC-II rose to the challenge magnificently. The logistics for the contingent were packed in 56 air and sea containers and it took a formidable effort of 13 sorties by super transporter AN-124 to transfer the ten helicopters, contingent equipment and self-sustainment assets.

The first batch of the IAC-II Contingent (Rotation-I) arrived at Bukavu on 26th Feb 2006 and became operational in March 2006. During Rotation-2, political negotiations with armed rebels continued at a frantic pace. As the scope of operations increased in the Eastern DRC, MONUC's requirement of air effort increased in the region near Tanganyika. A new detachment was necessitated at Kalemie which is located 218 nm south of Kavumu. Despite multiple constraints and a constant state of flux in the operational scenario, successive rotations continued to add their efforts to the building up of facilities and infrastructure for efficient operations.

Task MONUC is one of the largest and most robust peace keeping operations ever undertaken by the UN. The aviation contingents are by far the key elements in the arsenal of MONUC in achieving its mandate of maintaining peace in this huge country, the second largest in Africa. The tasks required by UN from the Mi-35 Sqn were to provide armed escort, fire support, anti-armour support, reconnaissance, information gathering and MEDEVAC, both by day as well as by night. The Mi-17 utility helicopters were to undertake tasks including troop induction, cargo carriage, underslung operations, SAR, CASEVAC and MEDEVAC, reconnaissance and surveillance, both by day and by night.

2005 Squadron It was in the month of Sep 2004 that the 'Firebirds' of 104 (H) Sqn, AF located in the hot yet serene environs of Rajasthan, learnt of the likely move to DRC. Equipped with upgraded Mi-35 helicopters, they were finally inducted in Feb 2005 to endow MONUC with the much needed night fighting capability. 104(H) Sqn, the oldest helicopter unit of the IAF was given this singular honour and thus became the first squadron of IAF to be deployed on foreign shores.

The Firebirds soon saw their first live action on 30th Jun 2005 in the Ituri Bde AOR, where they had to open fire to help extricate beleaguered

UN troops of Bangladesh and Pakistani battalions who had been ambushed by the militia. Since then, they were employed in more than a hundred operations all over the length and breadth of Eastern DRC, and always displayed a high degree of professionalism. The notable ones among these have been Op Songolo, where an offensive engagement by the squadron delivered a blow to the militia that had laid siege on a section of Pakistani APCs. This operation also saw the first ever night area domination mission being flown in DRC. In 'Op Falcon Sweep' during Jul-Aug 05, the Firebirds escorted more than ten Mi-17 helicopters to deploy troops for cordon and search operations against FDLR in South Kivu.

Op Eastern Eagle during Oct 2005 in Ituri region saw the squadron in action along with Guatemalan Special Forces and Uruguayan Riverine unit to address illegal arms smuggling through Lake Albert. Firebirds were also instrumental in getting rid of the FDLR militia operating in Virunga National Park as part of Op North Nexus. Through effective show of force and area domination over remote locales in the hinterland, they greatly assisted the peaceful conduct of the landmark elections in DRC during 2006. The most notable contribution from the Firebirds came during Op Sake in Nov 2006. In a swift move, a large number of disgruntled soldiers of the Govt army – FARDC, led by one of the most powerful warlords in the region, General Nkunda, took control of Sake, a town 25 km from Goma, and made deep inroads towards Goma. With the imminent threat of Goma falling to rebels, the Firebirds were tasked to spearhead the defence of this strategically critical town, the HQ of North Kivu Brigade. The ensuing operations saw the attack helicopters in action continuously by day and night for four days. In the face of a sustained fire by the rebels with small arms, mortars and RPGs, Firebirds successfully engaged the militia and their command centres with rockets and guns. This timely and effective attack crippled the rebel assault and resulted in securing Goma and regaining control of Sake. The Squadron had an enviable record of more than 3800 hrs of accident and incident free employment in DRC with nil fratricide and collateral damage. This enabled them to build an effective deterrence against the militia. It is said that Congolese militia were afraid of only two things – God and the Attack Helicopter. They had aptly nicknamed the attack helicopter 'Sura Mbaya' in Swahili language meaning 'The Ugly Face'.

The Kivu Hawks: 2006 (Helicopter) Squadron   In the last quarter of 2004, immediate and effective communication of MONUC forces to remote areas of DRC was the need of the hour. The inaccessible terrain with poor infrastructure of roads and non-existing communication lines only strengthened the demand for transportation by air. Hostile elements and presence of rebel groups precluded use of civilian effort. The flexibility of launching air effort by night was one of the pre-requisites and IAF did have this unique capability. In Feb 2005, six Mi-17 medium-lift, multi-role and night capable utility helicopters were inducted into Bukavu.

It was no surprise, that from the very first day, the 'Kivu Hawks' were at home! As they spread their wings over the verdant green hills of Bukavu, their reach and ubiquitous functionality was the cynosure of the entire Peace Keeping Force in the area. The flying tasks ranged from mundane passenger and VIP flights to exciting SHBO missions, search & rescues and air dominations. Within a week of induction into the area, a SHBO mission was flown over the stronghold of rebel forces. Op Falcon Sweep had thus signalled that they had come to roost for a long time. It also demonstrated a show of strength and resolve in the remote and inaccessible areas in the shortest time frames. It provided the long desired fillip to MONUC forces. Hawks had flown over 11500 sorties totalling more than 7000 hrs of accident free flying since induction in 2005. More than 63000 troops and passengers as well as 22,000,000 kg of cargo had been transported in the hostile environment of DRC. The active and continuous involvement of the Kivu Hawks in some of the most challenging operations such as Op South Sable, Op South Santlino, Op South Sailboard Op South Shrike, Op Kimia –II and Op Amani Leo have significantly facilitated the peace keeping ops and achievement of the overall aim of the MONUC forces.

Operations

The month of Jul 2005 was marked by one of the worst incident of human rights violations in South Kivu, where 47 women and children were burnt alive in confinement by militia. This was responded by UN by large scale induction of troops, aimed at instilling a feeling of security & confidence in locals. Accordingly, Op Mountain Viper was launched for area domination by inducting troops by air in stronghold area of the FDLR. The troops (Pak QRF & Guatemalan SF) were inducted & de-inducted the same day in a heliborne operation code named Op Falcon Sweep. This Op was a big success and the FDLR vacated their camp without resistance. The ground forces subsequently burnt down the camp to prevent re-occupation. In addition, Op Iron Fist was also launched by South Kivu Bde to flush out the Rwandan rebels from their strong hold.

Op Night Flash – II was re-launched during Aug 2005 by South Kivu Brigade to offset the growing influence of rebels. Another Op Safepath was launched along with FARDC with an aim to drive rebels from Kahuzibeiga Park. Op Bonne Nuit was launched during Sep 2005, wherein night static posts in major towns using APC were setup in various cross roads of the town as show of force and authority. Due to hectic arrangements towards smooth conduct of elections in the region, the patrols were positioned in various markets to monitor the ongoing activities. Op Night Flash–II was conducted in various parts of South Kivu sector wherein Pak & FARDC soldiers were tasked for a special night patrol for maintaining peace. Op Lake Watch was launched during Oct 2005 especially to focus towards patrolling along lake Kivu.

During Dec 2005, as a part of Op North Night Finale from Beni, Mi-35 helicopters were tasked to provide aerial support to IndBatt-II, which was engaged in flushing out militia in the area. The helicopters were guided by the ground elements to fire on certain camps, but the aircrew did not fire after assessing the nature of the camp, its activity level and other pertinent clues. It turned out to be a FARDC camp. This was avoided due to sound judgement of the aircrew.

During Jan 2006, both Mi-17s and Mi-35s took part in Op Paper Tiger in Bunia sector and Op Ituri Encourager. UN troops (Guatemalan SF) challenged the Ugandan rebels hiding along Sudan DRC border. A total of forty rockets were fired by Mi-35s against the militia. During this Op, eight Guatemalan troops were killed and five were seriously injured. Timely intervention of Mi-35s prevented further casualties. Op Ituri engager was carried out by PakBatt and BanBatt for area domination south of Bunia. 38 rockets were fired in two sorties. During this Op, one of the Mi-25 helicopters of IAC-1 was shot at resulting in critical hydraulic, fuel and air frame related emergencies. In addition, Mi-35s participated in Op Nut Shell against Gen Nkunda's rebels firing 60 rockets in North Kivu sector (90 Km north of Goma).

During Feb 2006, both Mi-17s and Mi-35s took part in Op South Sentinel in south Kivu sector. This was launched jointly by PakBatt and FARDC troops against the rebels (FDLR). A night medevac was carried out for a PakBatt soldier from Baraka to Bujumbura on 12th Feb. During March, both Mi-17 & Mi-35 helicopters took part in the Op South Sentinel and Op South Sustain in South Kivu sector– operating from Bukavu and Uvira. Both these were launched jointly by the Pak and FARDC troops against FDLR rebels. PakBatt continued its Op South stability in Bunyakiri sector and Op South Shuttle in Minembwe sector during May. On 29th May, it commenced another Op South Symbiosis in Kamanyola sector, wherein both Mi-17 and Mi-35 helicopters participated. During June, in view of the prevalent situation in the area, South Kivu Bde launched Op South Symbiosis (two Mi-17s & one Mi-35 participated in Walungu,Kanyola, Budodo & Nzimibra axis in two phases) on 4th June and 10th to17th Jun; South Stability from

26th May to 25th June in Buniyakiri / Hombo axis; South Stage (helicopters were tasked for lgs support and were on hot stand by for insertion/ extrication in Uvira & Kiliba sector); South Sail (helicopters participated in Uvira / Baraka axis in support of the Bde from 15th-18th Jun 2006); and South Synthesis from 26th-30th June in Walungu, Tubimbi and Mwenga sector. These operations were executed to protect the local population in the area.

The contingent participated in Op Smith Field, Op South Shuttle, Op Watershed-II & Op South Salvage during Oct 2006. During Sep 2008, Mi-17 helicopters inducted 104 Pak Batt troops in Numbi-Minova area towards pushing back rebel forces to the ceasefire line. The timely induction of the PakBatt troops assisted FARDC to blunt the offensive action by rebels. 120 rockets (57mm) were fired on rebel camps on the Masisi ridge. 84 rockets were fired during Oct 2008 by Mi-35s on targets identified to thwart any attempt by rebels to cross the ceasefire line. In one of the stray incidents, one of the Pak Batt soldiers fired at one local boy, who was trying to steal ration from the camp. The death of the boy led to lot of commotion and unrest among the civilians. Locals resorted to damaging UN property and vehicles. Additional troops were augmented to bring the situation under control.

Kamombo Op   A two aircraft detachment commenced its operation in support of Congo DRC government and South Kivu Bde at Kamombo from 23rd Jan 2009. This was necessitated due to abduction of two members of DDRRR by FRF– a rebel armed group in Kamombo. Negotiating teams and troops were transported in the area with a Mi-35 as escort. The abducted DDRRR personnel were finally released in Feb 2009. This Detachment was recalled on 14th Feb 2009.

FLYING EFFORT: IAF AV CONTG			
	MI-17	MI-25/35	ARMT
	SOR/ HRS	SOR/ HRS	57MM / 23 MM / 12.7MM
• IAC-1 :	15930 / 9590	5652 / 6170	1803 / – / 400
• IAC-2	13780 / 8680	4485 / 4720	297 / 100 / --
• IAC-UNMIS	22045 / 10445		
• TOTAL	51755 / 28715	10137 / 10890	2100 / 100 / 400

NO COLLATERAL DAMAGE

FOR A NOBLE CAUSE: GALLANTRY CITATIONS

Wing Commander Rakesh Kumar Negi (15869) F(P)

Wing Commander Rakesh Kumar Negi (15869) F(P) is on deputation to United Nations Peace Keeping Mission in Sierra Leone (UNAMSIL) as Commanding Officer of No.2000 (Helicopter) Squadron consisting of Mi-8s and Chetak. On 7th May 2000, Wg Cdr RK Negi, while flying over rebel held territory, monitored a distress call from another IAF Mi-8 that his aircraft has force landed and the rebels are approaching the aircraft from different directions. Realising the gravity of the situation, Wg Cdr RK Negi, displaying exceptional courage and presence of mind, immediately proceeded to the force landing site and successfully evacuated the 19 UN personnel and the crew before the rebels could capture them. The daring rescue mission received wide scale appreciation including from UN HQ, New York.

On 15th Jul 2000, he led his squadron in a joint air-land operation Op Khukri to extricate 222 beleaguered troops from the inhospitable terrain dominated by the Revolutionary United Front (RUF) rebels equipped with anti aircraft weaponry. Undaunted by the heavy odds, he courageously led his squadron from the front and flew 40 missions under extremely challenging terrain, weather and battle conditions. All these missions involved troop induction and redeployment from extremely restricted area, in treacherous weather with clouds touching trees tops and under continuous firing from rebels. Throughout the operation, he continued to inspire and motivate his crew by his personal example and was able to get the best out of them. He was instrumental in the swift success of Op Khukri.

Wing Commander Arun Tillu Samtani (17002) F(P)

Wing Commander Arun Tillu Samtani (172002) F(P) is the Commanding Officer of the Mi-35 Squadron of the IAF Contingent Sierra Leone-2000 for the United Nation Mission in Sierra Leone. 'Op Khukri' a joint Army Air Force Operation of the United Nations forces was executed to extricate 222 UN Peace Keepers held hostage for more than 75 days by Revolutionary United Front (RUF) rebels at Kailahun in Eastern Sierra Leone, Mi-35 helicopters of the IAF contingent were tasked to provide fire support to own troops at all times from dawn to dusk during this operation.

On 16th Jul 2000, he was tasked to carry out strike on RUF reinforcements coming through Manowa Ferry to impede the advance of UN troops heading from Kailahum to Daru and suppress the RUF fire at UN ground forces road convey. Undeterred by the enemy anti aircraft guns and heavy machine guns fire, he attacked enemy positions with rockets and front guns with pinpoint accuracy, inflicting heavy damage on rebel positions. Throughout the period of operation, he relentlessly carried out strikes at RUF positions with clinical accuracy at RUF, strongholds of Kailahaun, Giehun, Pendembu and Kuiva, which were located along the axis of advance of UN ground forces. He flew eight missions involving 14 hrs of flying in less than two days under extremely adverse weather and terrain conditions and under the threat of continuous firing by the rebels.

Squadron Leader George Thomas (17722) F(P)

Squadron Leader George Thomas (17722) F(P) is posted to Mi-8 flight of the IAF Contingent for the United Nations Mission in Sierra Leone. He is the Flight Commander of the Flight. Op Khukri a joint Army Air Force UN operation was conducted to extricate 222 Un peace keepers, held hostage for more than 75 days by Revolutionary United Front rebels in Kailahun. Mi-8 helicopters were tasked to carry out special heliborne operations in support of ground operations. These included insertion and extrication of troops into and from rebel strongholds. The helicopters were required to operate from unprepared sites and confirmed spaces under constant threat of enemy fire. Sqn Ldr George Thomas, a veteran helicopter pilots, was undeterred. Despite being fired upon on a number of occasions he led 24 missions into enemy territory, under extremely adverse weather conditions, over most inhospitable and thickly forested terrain. Throughout the conduct of this operation, the officer flew all the sorties assigned to him, with courage, enthusiasm and with utter disregard to personal safety.

On 8th Jun 2000, Sqn Ldr Thomas was tasked to escort a UN Convoy from Hastings to Rogberi, which was captured by rebels. His tasks included providing fire cover to the ground column. It involved flying the helicopter at extremely low altitudes over treacherous and thickly forested terrain and the threat of rebel fire. During this mission he spotted an ambush the rebels had laid for the convoy. Realising the gravity of the situation, the pilot engaged the enemy while at the same time informing the advancing column of the impending ambush. The rebels fired three rockets at the helicopter. With his exceptional flying skill, he successfully evaded these attacks. Undeterred, he held his position in the area and continued to guide the ground forces. This timely action thus prevented the ground forces from suffering heavy casualties due to enemy ambush.

Wing Commander Ajay Shukla (18793) F(P)

Wing Commander Ajay Shukla (18793) Flying (Pilot) is on the posted strength of Personnel Holding Section on deputation to the UN mission in

Congo since 27th Feb 2006. He is working as a squadron pilot in the Mi-35 Attack helicopter at Goma. On 25th Nov, a renegade warlord made deep advances towards Goma town with the town coming under an imminent threat of being taken over. Mi-35 attack helicopters were immediately pressed into action to prevent the fall of Goma. Leading from the front, Wg Cdr Ajay Shukla got airborne for fire support and reconnaissance in support of the Indian Brigade. In the repeated hostile fire from the rebels on his helicopter including that of rocket propelled grenades he remained undeterred and continued with the mission. Soon he detected heavy militia activity inside a well camouflaged camp and targeted with 57 mm rockets, leading to destruction of a key rebel command centre. This action by Wg Cdr Shukla delivered a crippling blow to the rebels and precipitated a severe disruption in their co-ordinated offensive. He flew a total of 10 crucial operational sorties, both by day and night. The sorties were carried out in very demanding conditions of hazardous terrain, inclement weather and enemy action. An experienced pilot on night vision goggles, Wg Cdr Shukla was also at the forefront of all night operations. He took calculated risks and displayed exceptional skills in handling the night targeting devices to effectively scan for movement of rebels on the critical nights of 26th and 27th Nov 2006. Crucial inputs provided by him exposed the game plan of the advancing rebels and enabled immediate friendly reinforcements to be sent. Throughout the operations, his accurate weapon delivery had a devastating effect on the vanguard rebel elements threatening Goma and on their camps being used for reinforcements, thereby preventing Goma town from falling into the hands of rebels.

CHAPTER XI

Laying a Firm Foundation: Training Helicopter Pilots

Introduction The key to a multi-skilled, adaptable and professional helicopter pilot in the field is sound training especially in the first few years. Theory and structured syllabi have to be mixed with useful on-the-job experience as a copilot if one has to fast-track the production of combat-ready pilots. In an age of very effective simulation technology, this experience can be ingrained without attendant risks and in a shorter time-frame. The IAF has a concrete vision of how it wants to train future helicopter pilots: basics on real aircrafts; skill building on a mix of real and simulation; applied flying in real terrain and environment interspersed with scenario training on simulators; mission rehearsals on simulators; and finally, recurrent training on emergencies, multiple workload, CRM etc on simulators. The service is acquiring Level-D full motion simulators and is also out sourcing some of its de-classified training.

This chapter highlights four institutions that are centres of excellence in their own fields – HTS, 112 HU, AEB and TACDE. All four have evolved constantly in keeping pace with what is happening all over the world and real requirements of the field. Helicopter assets being in demand forever, do not allow even these units to be free from the inevitable tasking for humanitarian and disaster relief.

Helicopter Training SchooL: The Alma Mater

Helicopter Training School (HTS), the alma mater of all helicopter pilots of the Indian Air Force, is located at Air Force Station Hakimpet, Secunderabad. The school also happens to be the largest helicopter unit of the Indian Air Force. Besides training pilots of the IAF and the Indian Army, it has also trained numerous pilots of friendly foreign countries; it is the only one of its kind in the entire country imparting both ab-initio and operational flying training on helicopters. The glorious past of the Helicopter Training Unit dates back to 2nd Apr 1962, when it was raised at Palam under Sqn Ldr SK Majumdar, it's first Commanding Officer. The school started functioning with a scanty strength of two trainees and a few Bell-47 helicopters in Jan 1963. Flying training initially was imparted on Bell-47 G3 and later on Chetak (Allouette III) helicopter from Oct 1973 onwards.

Just when the fledgling was finding its feet, and before it could settle down to actual business at Palam, the unit was moved to Allahabad on 30th Nov 1962 to function as a lodger unit of Pilot Training Establishment from 1st Dec 1962 and was re-designated as Logistic Support Training Unit (LSTU). It was to be first in a series of moves which were to follow. The unit moved in May 1967 from Allahabad to Jodhpur and stared functioning as a lodger unit of Air Force Flying College (AFFC) from 22nd May 1967 and was controlled by HQ Training Command through AFFC. It was at this location that the school was equipped with the new Chetak which continue to fly till date. On 1st Jun 1968, the LSTU was renamed as Helicopter Training School with Sqn Ldr HN Byrne as its First Commanding Officer. The school moved to its present location on 15th Oct 1973 under the command of Sqn Ldr KS Bindra and since then has continued to operate as a lodger unit of Air Force station Hakimpet. Rechristened as the Helicopter Training School in Jun 1968, it has come a long way since its humble beginnings to the present day where it stands as the epitome of excellence.

Role   The primary role has been to carry out ab-initio helicopter conversion training for pupil pilots of the Air Force, Army, Navy, paramilitary as well as foreign trainees. Apart from the fundamentals and nuances of rotor dynamics, trainees are kept abreast with the latest developments in helicopter aviation. Indeed, it has the distinction of being the largest helicopter unit in IAF. Since its inception, the school has made amazing progress over the years. The annual task, the strength of the staff and the number of trainees including Air Force and Army has expanded at unprecedented rates to cater to the vastly increased requirement of helicopter pilots in the Armed Forces. The school acquired international status in 1975 when trainees from Nepal, Bangladesh and Mauritius were also trained at this school. From a modest strength of two trainees in 1962, the school now trains nearly 100 trainees a year, with a peak monthly flying effort of up to 1000 hours, the largest of any helicopter unit in the Indian Armed Forces.

From 1971 onwards, courses conducted at HTS came to be known as Helicopter Conversion Course (HCC). The first HCC commenced training on 16th Jan 1971 with nine Army and five Air Force officers. Due to constraints on the availability of Cheetah helicopter, some of the Army officers were trained on the Chetak version. At present, all the training is done on Chetak helicopters. From Jan 1991, training for short service commission course was introduced. There was a change in trifurcation policy from Jan 2002 onwards for training of cadets at HTS. It was tasked for re-equipping of BFTS, Allahabad with Chetak helicopters for imparting flying training to army pilots in Jan 2006. Seven Chetak helicopters, 26 air warriors and associated tools/ ground equipment/ role equipment were successfully moved from HTS to BFTS. The school has trained more than 4000 helicopter pilots of Indian Air Force, Army, Navy, paramilitary and foreign officers. In addition to imparting professional flying training, ground subjects form an important part of the curriculum. Prior to commencement of flying, trainees have to undergo a knowledge building course on the helicopter (MCF). This is followed by regular lectures on professional, technical and service subject. Semester II (H) training also includes visits to other institutions like HAL, DRDL, MCEME and Survey of India. Active participation in group studies, presentation and outdoor activity is encouraged.

Operational Roles   Though the primary role is imparting basic flying and ground training, it has been regularly tasked with various operational and humanitarian missions. It has always been in the forefront for the service to the nation whether in war or in peace and was actively involved in operations right from its humble beginnings. Since its inception, it has pioneered and performed diverse roles in various parts of the country. In 1971, the unit went on a high alert, following developments in forward areas of north and east. It's involvement during hostilities though was restricted to manning detachments at forward bases in the north western sector, mainly for aerial reconnaissance and casualty evacuation prior to and during hostilities.

Relief Operations   The unit was on the lookout for some action and found a lots of it "down south" when it was called upon to provide succour and relief to the populace in Sri Lanka after a devastating Cyclone in Nov 1978. Two helicopters with ground crew flew out to Colombo on 25th Nov 1978, under the command of Sqn Ldr S Kalayanraman. Relief operations were undertaken from Batticaloa and Minueriya, with sorties starting at dawn on 27th Nov. 161 sorties logging 115 hours were flown to airlift 9800 lbs of food in addition to transportation of medical aid and passengers; a laudable effort, commended by the then Sri Lankan President, Mr J Jayawardane. HTS undertook aid of state administration in wake of the major earthquake that struck regions of Maharashtra and Karnataka (Latur) on 30th Sep 1993. The unit swung into action with a five helicopter detachment in support of massive relief operations and was instrumental in the successful disbursement of relief to the affected populace. Since its inception, the unit has pioneered and performed flood relief operations in various parts of the country especially the coastal regions of Tamil Nadu and Andhra Pradesh. It has been regularly tasked with various flood relief missions from time to time. The latest flood relief operations were in Sep-Oct 2009 which saw HTS rising up to the highest traditions of the Indian Air Force when devastating floods affected Andhra Pradesh and Karnataka. Tungabhadra River had broken its banks at many places. It responded with its characteristic lightening speed and promptly launched four aircraft, with the CO leading the contingent. The unique aspect of the operations was that National Highway-7 was used as the tarmac and operations carried out from the highway for the first two days when all bridges were broken and there was no rail or road connectivity in the flood affected areas.

While the helicopter fleet has extended yeomen service for more than half a century in the IAF, with ever increasing tasks and mission demands, the basic training exercises and syllabi had more or less been static. In view of the proposed modernization of the IAF, it was imperative to adopt an advanced and effective training programme in tune with the times and needs of the present and future. After an exhaustive study and analysis of many foreign air forces training syllabi and IAF's own experience of operational flying, the Commanding Officer, HTS brought out the need to expand the scope of exercises that were being taught at HTS which set the ball rolling in the evolution of an advanced training syllabus. The advance syllabus was an expansion of existing and incorporation of newer manoeuvres after exhaustive discussion and deliberation by a dedicated team. The current training program at HTS is oriented towards operational flying which is more practical in field and combat situations.

Vignettes of Service to the Nation

Medak Rescue On 14th September 2008 at 1430 hrs, the CO received a message from HQ Training Command that ten people were marooned due to flash floods caused by the release of water from the Singur Reservoir and only an immediate rescue could save their lives. The approximate location of the site was 20 nm west of Medak. The crew consisting of Isser, Dongre and Kulkarni (Winch Operator) got airborne at 1515 h from Hakimpet, just 45 minutes from the first warning (on a Sunday). The helicopter proceeded to carry out a search along the Manjra River towards Nizamsagar reservoir. After a brief search, a small temple was located surrounded by raging flood waters on the other side of the river opposite the famous Bhavani temple. There were ten people stranded on a rooftop (terrace) awaiting rescue and hundreds of people had gathered on the other river bank.

During the high/low recce of the site it was observed that the site was surrounded by tall trees and numerous wires/poles. The clear area was barely 15 ft × 15 ft making landing impossible. The aircraft was brought to OGE hover of 30 ft and the rescue cable was lowered by the winch operator. Since the people involved were untrained and panicky civilians they were not able to secure themselves to the rescue strop and damaged it in the process. The rescue strop was then retrieved and the aircraft descended to a height of ten ft over the terrace with just the bare minimum clearance from the obstructions. This was done in view of the rising river and limited time available to rescue before the flood waters took their toll. Thereafter, one by one the survivors were winched (they simply grabbed and hung on to the cable) up into the aircraft and hauled in by the co-pilot and the winch operator while the captain maintained a steady hover despite gusty winds. After every two survivors were picked up, the helicopter proceeded to a paddy field across the river and dropped them off. In all ten survivors were lifted to safety in five shuttles. The total sortie duration was 1:45 hrs non-stop without landing. Within a few hours, the then Chief Minister of Andhra Pradesh, Dr YSR Reddy complimented the helicopter unit personnel of IAF, Hakimpet for rescuing the ten lives. The Commissioner for Disaster Relief specially thanked the IAF for the near immediate and timely response. In his letter to the AOC-in-C, he commended the IAF for this act of courage and professionalism, which the people of Andhra would remember always.

Hovering Angels of Kurnool   Unprecedented rains in North Karnataka from 29th Sep 2009 onwards flooded several districts and cut off areas like Bijapur, Bagalkot and Bellary, some places receiving over 50 cm of rains in a single day. The floods inundated the Krishna and Tungabhadra rivers.

Karnataka released 25 lakh cusec of water from Almatti and Narayanpur dams in a single day, a record of sorts. With Andhra Pradesh itself in a grip of severe rains in the last week of September, all reservoirs were brimming to their capacities. On 1st October, reports were received that water levels were rising dangerously in the town of Mantralayam. The water levels rose so high that it submerged most of the villages inhabited as far as one km from the normal river bank. That condition of these villages was critical is an understatement.

The water level was so high that a village which was about one km away from the river bank was completely under water. The nearest bank was too far away now and the water level was as high as roof tops. With water levels still rising and most lives hanging around on these rooftops, villagers were not sure whether they were going to survive the next hour or not. To their rescue came the helicopters of the Indian Air Force. Four helicopters of HTS were tasked to provide relief and rescue operations to the people of flood affected parts of Andhra Pradesh and Karnataka. On 2nd of October, the task was challenging as operating conditions were critical with villagers in a dire state of survival. Marginal weather made the task even more challenging. The flood relief and rescue missions were carried out with a zeal and passion rarely seen, yet in a most professional manner by the air warriors of the IAF.

To add to the tough conditions, the fuel bowser was unable to reach Kurnool due to the breaks in the road bridges over Krishna and Tungabhadra. On one side the fuel bowser was stuck 20 nm short of Kurnool town and on the other the fuel requirement was mandatory to continue any type of rescue/relief operations. Analysing the gravity of situation, the Task Force Commander got airborne and on assessment of the situation realised that there was no clear dry ground available to allow the bowser to approach without getting bogged down owing to rains. Keeping in mind the Flight Safety aspects, a landing spot on the national highway was selected and a decision to operate from the NH-7 taken. The rescue and relief missions continued. A total of eight rescue missions saved 47 lives that would have surely perished. Each mission was a daunting and challenging experience in itself. The missions varied from picking up people from roof tops to those surviving on trees. In one case, a total of nine lives were stuck on a roof top. The increase in water flow washed away half of the house along with four persons, leaving behind the remnants with remaining five persons hanging on between life and death. To their fortune, and within no time, a Chetak helicopter came as a saviour. All of them were winched up in the nick of time as per the District Collector's report.

In another case, a tribal family of four consisting of husband, wife and two small children aged between four and six each were stuck on a tree in the centre of the fast flowing river. It was not an easy task to rescue these people from the centre of tree with all its branches spread around. No orders or procedure exists that specified the manner in which such a mission could be undertaken. The crew of the Chetak helicopter used their ingenuity and experience and rescued the entire family in a most professional yet flexible approach. After their rescue, the crew noticed that their condition was very critical and their skins were totally parched and coming out in flakes. They had been stuck in that tree for over 72 hours without food, water and sleep. Every such mission makes helicopter pilots the world over feel proud of their machines.

Many relief missions were carried out during the seven-day ordeal. A total of 1,20,000 kgs of relief material including water, food and medicines were dropped to the victims who were in dire need. People were stuck on roof tops and the small Chetak helicopter was able to carry out drops accurately roof-to-roof. It was a difficult task as the area to drop food/water was very small and in the centre of a furious and fast flowing river. The IAF carried out these operations in a most accurate and efficient manner. These relief material were not only a must for their physiological needs but more importantly to generate in the victims a will to survive and a ray of hope to live another day. All missions were efficiently carried out by seven helicopters of the IAF. Each mission, whether it was a flood relief operation or a rescue mission, was a daunting task taken on with professionalism. The entire mission can rightly be summed up in the words of District Collector of Kurnool in the press conference on 7th Oct 2009 where he stated "The people of Kurnool will always be indebted forever to the efforts of helicopters of the IAF. Not only did the IAF save 47 lives, but helped 6000-7000 people every day to believe that survival and help was just around the corner".

Highway converted into air base

Authorities identify 18 villages for air dropping and rescuing stranded persons

Effective Teamwork and Courageous Effort Saved Many Lives

D.O. No. 4835 D V/2009

रक्षा मंत्री
भारत
**MINISTER OF DEFENCE
INDIA**

20th Oct, 09

Dear Air Chief Marshal Naik,

This is to place on record my appreciation for the commendable job done by the Armed Forces personnel during the relief work in Karnataka and Andhra Pradesh as both the States were devasted by the recent unprecedented floods. Air Force has once again proved its readiness not only in defending the country but also safeguarding the lives and property of civilian society during national calamity.

Please accept my heartiest congratulations and convey the same to all the personnel posted on the relief work.

With best wishes,

Yours sincerely,

(A.K. Antony)

**Air Chief Marshal PV Naik
PVSM   VSM   ADC**
Chief of the Air Staff
Air Headquarters

Letter of appreciation by Raksha Mantri

Twin-Engine/Medium Lift Helicopter Conversion

112 Helicopter Unit, popularly known as 'Thoroughbreds', was raised in Jorhat on 1st Aug 1963, it was then equipped with Bell 47G helicopters with a task of logistics support, air and land rescue, communication and Casevac for the remote and inaccessible areas of the then NEFA, now Arunachal Pradesh. It moved to Bagdogra in Nov 1966 and was re-equipped with Chetak and Cheetah helicopters where it flew innumerable rescue missions and provided operational support to the Indian Army from sea level to the altitudes in excess of 20,000 feet.

It proved its mettle in the 1971 war, transporting men and material in to the war zone. After 20 years in the east, on 11th Jul 1982 the unit was re-equipped with the rugged Mi-8 helicopters and moved to its present location in Yelahanka, Bangalore. Since then it has been tasked in Operation Pawan and Operation Cactus. In the same breath, the unit has also been saving lives and providing relief whenever tasked during the times of distress like floods, cyclones and in particular the infamous Asian Tsunami.

This unit was given the task of conducting ad-hoc Mi-8 conversion courses for pilots in Jul 1984. The basic flight engineers course was started in Aug 1984. Conversion for pilots on Mi-8, after their ab-initio Semester I helicopter training at HTS, was commenced in Dec 1990. 112 HU was granted Flying Training School status from Oct 1992. The unit also conducts RHSCP course for pilots detailed by Air HQ. The Thoroughbreds are the only Flying Training Establishment of the Indian Air Force with a dedicated war time role for which the unit regularly undertakes operational missions. It has the distinction of being adjudged the "Best helicopter Unit of IAF" for the period 1998-99.

TACDE and Helicopter Tactics

Air Power has emerged as the dominant arm of a nation's military power. Over the decades, advances in science have revolutionised Air Power and as a result, it has become highly technology sensitive. Notwithstanding this, it is fairly apparent that more than technology, it is tactics that provide the cutting edge in air warfare. At the time of formation, inherited tactics and doctrine were initially utilised by the IAF. In time, a need arose for an independent establishment where fighter aircrew could evolve tactics for the effective employment of air assets. Wars in quick succession post independence, added urgency to this requirement.

On 1st Feb 1971, Tactics and Combat Development and Training School (T&C D&TS) was established at Adampur. This institution was soon baptised by fire when it was tasked to develop tactics for Counter Air Operations by night in the 1971 war. The surprise effects of these tactics contributed significantly to the IAF seizing the initiative at the very onset of the war. The Institution then moved to Ambala and a short while later, to Jamnagar. In 1972, the T&C D&TS was redesigned Tactics and Air Combat Development Establishment (TACDE). Initially, a succession of Fighter Combat Leaders' (FCL) Courses was conducted. Subsequently, TACDE expanded to conduct courses for Fighter Strike Leaders, Pilot Attack Instructors, Helicopter Strike Leaders, Master Fighter Controllers, Tactical Fighter Controllers and SAGW crew. TACDE also has a Development Wing which has constantly sought to remain abreast of contemporary tactics. This has enabled the IAF to keep pace with fast-changing developments in the field of military aviation. Through regular interactions with field formations, TACDE has also optimised the use of air assets and been able to foster greater synergy through the ambit of IAF operations.

The Vietnam War added a new dimension to air warfare with helicopters finding a number of new roles in the battle field. There was also the realisation of the limitations of helicopters as a weapon platform due to their vulnerability. With the importance of helicopters as a versatile weapon platform being realised the world over, the IAF felt the need to enhance their combat potential. To fulfil this requirement, TACDE was tasked to carry out trials on Chetak and Mi-8 helicopters in the early eighties. The first dedicated Helicopter Operation Orientation Capsule was conducted at TACDE from 18th January 1985 to 31st January 1985. Thereafter, it was actively involved in formulating SOPs and combat tactics for all types of helicopters in the IAF. As early as 1990, realising the ever increasing role of helicopters in the modern day battlefield, it proposed the conduct of dedicated Helicopter Combat Leader Courses.

Finally, the mantle of TACDE enveloped Helicopter Combat Employment as well with the commencement of first Helicopter Combat Leader's Course in June 1997. There were numerous problems that had to be tackled, since there were no Directing Staff from helicopters. Soon Wg Cdr Bharali and Sqn Ldr Maheshwar were posted in as Helicopter Directing Staff. Nirmal Rai and Jaideep Hora were co-opted for the course. The conduct of the course was a major challenge to TACDE, since there was a wide gap between the perceptions of seasoned helicopter pilots and newer proponents of helicopter tactics. The main aim of the HCL course was to optimise the use of this relatively unexploited machine and teach pilots the fine art of instructional technique in air combat and air to ground weapons delivery. The course curriculum, besides encompassing all exercises related to airborne weapon delivery also included a comprehensive and exhaustive ground subject syllabus. They also developed manoeuvring and situational awareness skills alongside other helicopter specialist roles. It also included a comprehensive helicopter versus fighter phase. The pilots on completion of the course graduated as Helicopter Combat Leaders and were responsible for combat orientation and training at squadron/unit level.

Objectives of the Course

- Development of leadership skills.
- Progress the officers to levels of operational preparedness enabling them to train pilots in the field in keeping with the high standards desired by the IAF.
- Refinement of tactical and combat flying.
- Build up situational awareness and ability to carry out critical analysis.
- Enhance ability to plan operational solutions and integration of various elements in larger force packages.
- Emphasis on role oriented training.

Initially the helicopter tactics were drawn out of the existing fighter tactics. However, as the courses progressed, these were modified to cater to the peculiarities of the machine and evolved into type specific tactics. Presently TACDE has two helicopter DS on its posted strength. The establishment is engaged in continuously evaluating and updating tactics based on the ever changing threat perceptions and the environment in the TBA. In an endeavour to reach out to the field and create wider

awareness among the helicopter crew the establishment started conducting Helicopter Combat Employment Capsule (now known as Helicopter Operational Employability Capsule) in each Command. Pilots from local units are exposed to various facets of the helicopters in operational roles and are also introduced to Basic Helicopter Manoeuvring for offensive and defensive roles. TACDE also provides the opportunity for helicopter aircrew to operate in simulated TBA environment during the AKRAMAN phase of FCL/FSL course. During this phase, helicopter pilots are exposed to the SAGW threat and interception by fighter aircraft preparing them in as realistic an environment as can be generated during peacetime.

The badge has a background of the rising sun signifying the emergence of the helicopter as a formidable weapon platform. The helicopter enroute a radar strike mission is depicted as turning to face the intercepting fighters. The radar in black, at the bottom left signifies the ability of helicopters to fight the fighters and protect themselves. The orange background depicts strike at dawn/dusk (orange sky). This is symbolic of the confidence and courage of a pilot, well-trained and skilled in the art of air combat, as also of the lethality and versatility of the helicopter. The Motto **'Learn to Lead, Lead to Attack, Attack to Destroy'** embodies the display of leadership qualities imbibed during the course, to attack and destroy the target each and every time. The photograph below is of the first HCL Course.

The Kurnool - Mehboobnagar Saga: 30 Sep - 08 Oct 2009

Aircrew Examining Board

The Aircrew Examining Board (AEB) was established at Palam in Jul 1951. It was originally called as the Aircrew Training and Testing Team. On 24th Mar 1956 it was renamed as the AEB. The Board moved to its present location at Hindan in Mar 1965. To enable uninhibited functioning, it is established as an independent authority and is under the functional control of the Director General (Inspection & Safety). The aim of AEB is to maintain a high standard of professional knowledge and skill of aircrew. It also functions as a 'Quality Audit' team to provide guidance to commanders for assessing suitability of aircrew to undertake specific missions and career courses. The Board achieves this by carrying out categorisation, instrument rating and standardisation of aircrew. *Dhee Kaushalam Poornata* (Knowledge, Merit and Perfection) is the befitting motto of this Board which is committed to ensure excellence in the field of aviation.

In May 1969, Delta Flight was added to the Board with the responsibility of inspecting the helicopter stream. It undertakes all Categorisation and Instrument Rating tests of helicopter pilots of field units once a year, standardisation of trainee helicopter pilots in FTEs twice a year and standardisation of under training Instructors of helicopter stream at the Flying Instructor School, twice a year. It also conducts Instrument Rating Instructors & Examiners courses for helicopter pilots once every year. Three to four Air Force Examiners (AFEs) handpicked from the fleet undertake these tasks and inspect all the helicopter field units/squadrons and FTEs. The aim of Delta Flight is to set and maintain high standards of professional knowledge and skill amongst helicopter aircrew, and make them conscious about flight safety and motivate them to become accomplished helicopter pilots. With the induction of state of the art helicopters today, the AEB stands as a venerable institution, shouldering a wide variety of responsibilities and accomplishing multifaceted tasks. In the years to come, it is destined to soar to even greater heights in continuing to live up to the IAF motto of *Touch the Sky with Glory*.

Often referred to as 'Weekend Husbands', the helicopter Air Force Examiners spend almost three weeks away from home every month, inspecting helicopter units/ squadrons stretching from Leh in the North to Carnic in the far South East and Mohanbari in the East to Jamnagar in the West. Approximately four to six hundred operational helicopter pilots are categorized every year, besides standardisation of 200 ab-initio and applied stage trainees.

Epilogue

As I have written in the beginning, this project was a humbling experience. Like the larger legacy of the Indian Air Force, the professional ethos and pragmatism of the fleet truly stood out. Yet, gallant pilots and aircrew of the IAF have achieved more than what was mathematically possible because of that added quality that is the true flavour of the service – a passion for excellence and a can-do attitude.

This book hopefully provides a window to this legacy, which awaits more research and recounting by air warriors who are serving and those that have retired. There are challenges galore in the future – rapid modernisation and transformation brought about by visionary leaders of the service, automation and unmanned technology which is growing in leaps and bounds, training for the right skills and attitudes etc. A knowledge of what our forebears in the IAF have done will only point out the correct direction to move.

The more likely conflicts will be limited in scope, time and area – more akin to swift special operations. The response time of the IAF, when forewarned and kept in the planning loop, has always been good and timely. The problem starts when any service is kept out of the conceptual phase and brought in only when the chips are really down – as in Kargil in 1999. Fourth generation warfare requires great adaptability, flexibility in thinking intellect to understand finer nuances of complex problems. Similarly, humanitarian assistance and disaster relief are complex issues where the civilian authorities' point of view needs to be understood. Today, any situation involving the two scenarios, and many of those between the spectrums, will always involve helicopters, UAVs and the foot-soldier. The partnership needs to grow as shown by earlier campaigns and operations that have been flagged in this book. Rather than going for turf wars, there is a need to respect core-competencies. Such expertise needs decades of institutional memory, evolution and management of change.

Is there a need to change the successful model of helicopter ownership and integration that the country chose in 1986? I do not think so – the unblemished record carries the answer. The real question is can a country like India afford this cost– when slight tweaking can give us better results? Whatever decision the government makes, the legacy of the IAF's helicopter field will always stand tall to inspire future air warriors.

Honours and Awards

President's Colours/Standards

In military organisations, the practice of carrying colours or standards acts as a rallying point for troops and to mark the location of the commander. This is thought to have originated in ancient Egypt about 5000 years ago. It was formalised in the armies of medieval Europe, with standards being emblazoned with the commander's coat of arms. As armies became trained and adopted set formations, each battalion's/regiment's ability to maintain its formation in the heat and dust of battle was critical to its success. Due to the advent of modern weapons, and subsequent changes in tactics, Colours are no longer carried into battle, but continue to be used at events of formal gatherings.

In the present times, regimental flags/colours are awarded to a fighting unit by a head of a state during a ceremony. They are, therefore, treated with reverence as they represent the honour and traditions of the unit. In the Indian armed forces, the award of the President's Colours/Standards upon a Unit or Squadron is one of the greatest honours bestowed in recognition of exceptional service rendered by it to the nation, both during peace and war. The following Helicopter Squadrons and Units have been presented with the President's Standards/ Colours.

104 Helicopter Sqn: 08 Nov 1988

114 Helicopter Unit: 13 Nov 1996

109 Helicopter Unit: 02 Mar 2009

105 Helicopter Unit: 25 Mar 2009

107 Helicopter Unit: 09 Mar 2010

110 Helicopter Unit: 26 Mar 2011

111 Helicopter Unit: 06 Mar 2012

Presidents Standard Presentation

After being formed at Srinagar on 1st June 1960, the unit relocated to Bareily in 1965 and further to present location at Jodhpur in 1971. Since its inception, the unit has flown a large variety of helicopters , namely Sirkorsky-62, Bell-47 6, Chetak, Mi-4, Mi-8 and now the versatile Mi-17, across inhospitable terrain spanning Leh to Antarctica and the Thar to the plains of central India.

107 HU & 104 HU Being honoured by the President

Presentation of President's Standard
107 HU Air Force

C/O 56 APO
09 MAR 2010

Desert Hawks

ISSUED BY THE ARMY POSTAL SERVICE - 23/2010

155 HELICOPTER UNIT
INDIAN AIR FORCE

C/O 56 APO
25 MAR 2009

Presentation of President's Standard

105 HU, AF
1959-2008

ISSUED BY THE ARMY POSTAL SERVICE - 09/2009

AWARD OF PRESIDENT'S STANDARD

Vanguards

GALLANTRY AWARDS: IAF HELICOPTER AIRCREW

MAHA VIR CHAKRA

SI No	Ser No	Rank	Name	Branch/Trade	Year	Award Date	Gazette Notification Number
1	3460	GP CAPT	C SINGH AVSM VrC	F(P)	1972	20-Jan-72	20-PRES/72 DT 12 FEB 72

KIRTI CHAKRA

SI No	Ser No	Rank	Name	Branch/Trade	Year	Award Date	Gazette Notification Number
1	11607	WG CDR	MAHESHWAR DUTT	F(P)	1988	26-Jan-88	26-PRES/88 DT 2 APR 88
2	17870	WG CDR	BPS KRISHNA KUMAR	F(P)	2005	15-Aug-05	125-PRES/2005 DT26 NOV 05
3	17717	WG CDR	SUDHIR KUMAR SHARMA	F(P)	2005	26-Jan-05	34-PRES/2005 DT 9 APR 05

VIR CHAKRA

SI No	Ser No	Rank	Name	Branch/Trade	Year	Award Date	Gazette Notification Number
146	3950	SQN LDR	AS WILLIAMS	GD(P)	1962	12-Nov-62	AFO 211/63
147	5053	FLT LT	KL NARAYANAN	GD(P)	1962	12-Nov-62	AFO 211/63
148	3973	SQN LDR	SK BADHWAR	GD(P)	1962	12-Nov-62	AFO 211/63
149	3460	SQN LDR	C SINGH AVSM	GD(P)	1963	26-Jan-63	18-PRES/63 DT 16 FEB 63
150	4436	FLT LT	KK SAINI	GD(P)	1964	1-Jan-64	3-PRES/64 DT 11 JAN 64
151	6893	FLT LT	CM SINGLA	F(P)	1972	18-Jul-72	92-PRES/72 DT 29 JUL 72
152	6892	FLT LT	PK VAID	F(P)	1972	18-Jul-72	92-PRES/72 DT 29 JUL 72
153	8378	FLT LT	MP PREMI	F(P)	1972	18-Jul-72	92-PRES/72 DT 29 JUL 72
154	5591	SQN LDR	CS SANDHU	F(P)	1972	18-Jul-72	92-PRES/72 DT 29 JUL 72
155	10948	FG OFFR	S RAMESH	F(P)	1972	18-Jul-72	92-PRES/72 DT 29 JUL 72

Sl No	Ser No	Rank	Name	Branch/Trade	Year	Award Date	Gazette Notification Number
156	11378	FG OFFR	S DHILLON	F(P)	1972	18-Jul-72	92-PRES/72 DT 29 JUL 72
157	9834	FLT LT	PR JAMASJI	F(P)	1972	18-Jul-72	92-PRES/72 DT 29 JUL 72
158	5863	SQN LDR	SK CHOUDHURY	F(P)	1972	18-Jul-72	92-PRES/72 DT 29 JUL 72
159	11887	WG CDR	DR DURAISAMI	F(P)	1988	15-Aug-88	37-PRES/88 DT 30 APR 88
160	17149	FLT LT	V PRAKASH	F(P)	1988	15-Aug-88	37-PRES/88 DT 30 APR 88
161	11438	SQN LDR	TK VINAY RAJ	F(P)	1988	15-Aug-88	37-PRES/88 DT 30 APR 88
162	14287	SQN LDR	RAJBIR SINGH	F(P)	1988	15-Aug-88	37-PRES/88 DT 30 APR 88
163	11285	WG CDR	VN SAPRE	F(P)	1988	15-Aug-88	37-PRES/88 DT 30 APR 88
164	16077	FLT LT	AN HANFEE	F(P)	1989	26-Jan-89	23-PRES/89 DT 8 APR 89
165	17449	FLT LT	NM SAMUEL	F(P)	1989	26-Jan-89	23-PRES/89 DT 8 APR 89
166	11336	WG CDR	CD UPADHYAY	F(P)	1989	26-Jan-89	23-PRES/89 DT 8 APR 89
167	16983	FLT LT	M SINGH	F(P)	1990	26-Jan-90	33-PRES/90 DT 28 APR 90
168	16794	FLT LT	ATANU GURU	F(P)	1990	26-Jan-90	33-PRES/90 DT 28 APR 90
169	12032	WG CDR	KK YADAV	F(P)	1990	26-Jan-90	33-PRES/90 DT 28 APR 90
170	11590	WG CDR	DN SAHAE	F(P)	1990	26-Jan-90	33-PRES/90 DT 28 APR 90
171	21545	FLT LT	BHUPINDER	F(P)	1998	15-Aug-98	113-PRES/98 DT 23 JAN 99
172	16074	WG CDR	ANIL KUMAR SINHA	F(P)	1999	15-Aug-99	18-PRES/2000 DT 11 MAR 2000

SHAURYA CHAKRA

Sl No	Ser No	Rank	Name	Branch/Trade	Year	Award Date	Gazette Notification Number
1	7676	FLT LT	RN PANDEYA	F(P)	1968	26-Jan-68	26-PRES/68 DT 6 APR 68
2	7741	SQN LDR	RPS DHILLON	F(P)	1975	1-Jul-75	34-PRES/76 DT 1 MAY 76
3	6756	SQN LDR	MS SEKHON VrC VM	F(P)	1976	26-Jan-76	40-PRES/76 DT 29 MAY 76
4	9031	FLT LT	RC GHILDIYAL (POSTH)	F(P)	1978	26-Jan-78	14-PRES/78 DT 8 APR 78
5	9071	SQN LDR	KS PARIHAR	F(P)	1978	26-Jan-78	14-PRES/78 DT 8 APR 78
6	10117	SQN LDR	SS SIDHU VM	F(P)	1978	26-Jan-78	14-PRES/78 DT 8 APR 78
7	7441	SQN LDR	FJ WILLIAMS	F(P)	1979	26-Jan-79	16-PRES/79 DT 28 APR 79
8	8762	SQN LDR	GS BRAICH	F(P)	1980	26-Jan-80	37-PRES/80 DT 24 MAY 80
9	9763	SQN LDR	MP RANE	F(P)	1980	26-Jan-80	37-PRES/80 DT 24 MAY 80
10	8136	SQN LDR	RANBIR SINGH CHAUHAN	F(P)	1981	26-Jan-81	28-PRES/81 DT 18 APR 81

11	13602	FLT LT	PL NAVALE	F(P)	1982	26-Jan-82	19-PRES/82 DT 10 APR 82
12	15558	FG OFFR	J BALASUBRAMANIAM	F(P)	1982	26-Jan-82	19-PRES/82 DT 10 APR 82
13	10452	SQN LDR	GURCHARAN SINGH MADAN	F(P)	1981	2-Apr-82	20-PRES/82 DT 10 APR 82
14	7018	WG CDR	BS CHHOKER	F(P)	1982	23-Dec-82	54-PRES/82 DT 1 JAN 83
15	11316	SQN LDR	SK BHATIA	F(P)	1983	26-Jan-83	14-PRES/83 DT 19 MAR 84
16	13573	FLT LT	BD SINGH	F(P)	1983	26-Jan-83	14-PRES/83 DT 19 MAR 84
17	10920	SQN LDR	BN RAMACHANDRA	F(P)	1984	26-Jan-84	25-PRES/84 DT 24 MAR 84
18	11023	SQN LDR	KPN SINGH	F(P)	1984	26-Jan-84	25-PRES/84 DT 24 MAR 84
19	12407	SQN LDR	SK DIXIT	F(P)	1984	26-Jan-84	25-PRES/84 DT 24 MAR 84
20	7017	WG CDR	KK SANGAR VM	F(P)	1985	26-Jan-85	23-PRES/85 DT 16 MAR 85
21	10462	WG CDR	V NATARAJAN	F(P)	1985	26-Jan-85	23-PRES/85 DT 16 MAR 85
22	11290	SQN LDR	TPS CHHATWAL VM	F(P)	1985	26-Jan-85	23-PRES/85 DT 16 MAR 85
23	10561	WG CDR	CM RAO (POSTHUMOUS)	F(P)	1990	26-Jan-90	35-PRES/90 DT 28 APR 90
24	14094	SQN LDR	S MISHRA	F(P)	1990	26-Jan-90	35-PRES/90 DT 28 APR 90
25	11442	GP CAPT	FH MAJOR VM	F(P)	1993	26-Jan-93	66-PRES/93 DT 8 MAY 93
26	12957	WG CDR	S CHANDER	F(P)	1993	26-Jan-93	66-PRES/93 DT 8 MAY 93
27	12954	WG CDR	VK ARORA	F(P)	1994	26-Jan-94	25-PRES/94 DT 9 APR 94
28	17007	SQN LDR	AK GUPTA	F(P)	1994	26-Jan-94	25-PRES/94 DT 9 APR 94
29	661178	SGT	RAMESH CHAND	FLT GNR	1994	26-Jan-94	25-PRES/94 DT 9 APR 94
30	17697	SQN LDR	TP NANDA	F(P)	1995	15-Aug-95	182-PRES/95 DT 7 OCT 95
31	13812	WG CDR	AS SLAICH VM	F(P)	1996	26-Jan-96	16-PRES/96 DT 9 MAR 96
32	14281	WG CDR	HPS NATT VM	F(P)	1996	26-Jan-96	16-PRES/96 DT 9 MAR 96
33	15434	WG CDR	RS MANN	F(P)	1996	26-Jan-96	16-PRES/96 DT 9 MAR 96
34	19534	FLT LT	SANDEEP JAIN (POSTH)	F(P)	1997	15-Aug-97	84-PRES/97 DT 1 NOV 97
35	23546	PLT OFFR	V BHAGWAT (POSTH)	F(P)	1998	26-Jan-98	32-PRES/98 DT 18 APR 98
36	17012	SQN LDR	MSM NASSAR	F(P)	1998	15-Aug-98	114-PRES/98 DT 23 JAN 99
37	23771	FLT LT	SR KAGDI (POSTH)	F(P)	2001	15-Aug-01	130-PRES/2001 DT 9 FEB 02
38	15863	GP CAPT	VV BANDOPADHAY VM	F(P)	2005	15-Aug-05	126-PRES/2005 DT 26 NOV 05
39	25858	FLT LT	ARISETTI VIJAY KUMAR	F(P)	2005	15-Aug-05	126-PRES/2005 DT 26 NOV 05
40	22947	SQN LDR	SHANTANU BASU (POSTH)	F(P)	2008	26-Jan-08	25-PRES/2008 DT 10 MAY 08
41	25603	SQN LDR	PRANAY KUMAR	F(P)	2008	15-Aug-08	106-PRES DT 08 NOV 08
42	25871	SQN LDR	TK CHAUDHRI	F(P)	2009	15-Aug-09	114-PRES/2009 DT 05 DEC 09
43	776951	SGT	MUSTAFA ALI (POSTH)	FLT ENG	2009	15-Aug-09	114-PRES/2009 DT 05 DEC 09
44	23569	WG CDR	FELIX PATRICK PINTO	F(P)	2012	26-Jan-12	31-PRES/2012 DT 24 MAR 12

YUDH SEVA MEDAL

Sl. No.	Rank	Name	Per No.	Branch	Year	Date of Award
1	FG OFFR	AP SRIVASTAVA	17471	F (P)	1988	15 Apr 88
2	FG OFFR	AJAY MASSAND	17730	F (P)	1988	15 Apr 88
3	FLT LT	VS BHARTI	16611	F (P)	1989	26 Jan 89
4	SQN LDR	KJ SINGH	13799	F (P)	1989	26 Jan 89
5	FLT LT	A MOKASHI	17473	F (P)	1990	26 Jan 90
6	FLT LT	N TANWAR	17461	F (P)	1990	26 Jan 90
7	FLT LT	SANJAY MITTAL	16773	F (P)	1990	26 Jan 90
8	WG CDR	VS RANAWAT	12390	F (P)	1990	26 Jan 90
9	WG CDR	RV KUMAR	10918	F (P)	1990	26 Jan 90
10	FLT LT	AS GILL	17317	F (P)	1990	26 Jan 90
11	FLT LT	P GUPTA	17349	F (P)	1992	26 Jan 92
12	WG CDR	DD MANDPE	13937	F (P)	1995	26 Jan 95
13	WG CDR	NK UPADHYAY	12779	F (P)	1995	26 Jan 95
14	SQN LDR	PE PATANGE	17706	F (P)	1995	26 Jan 95
15	GP CAPT	BS SIWACH AVSM VM	12413	F (P)	2001	26 Jan 01
16	WG CDR	BIBHUTI KUMAR	18495	AE (M)	2005	26 Jan 05
17	GP CAPT	KULWANT SINGH GILL VM	15520	F (P)	2005	26 Jan 05
18	WG CDR	SUNIL KUMAR	17713	F (P)	2005	26 Jan 05
19	GP CAPT	ATANU GURU VR C	16794	F (P)	2008	26 Jan 08

VAYU SENA MEDAL (GALLANTRY)

Sl No	Ser No	Rank	Name	Branch/Trade	Year	Award Date	Gazette Notification Number
1	4025	SQN LDR	A DALAYA	GD(P)	1966	1-Jan-66	14-PRES/66 DT 12 FEB 66
2	4440	FLT LT	BS JASWAL	GD(P)	1963	26-Jan-63	10-PRES/63 DT 26 JAN 63
3	5181	FLT LT	SS SODHI	GD(P)	1963	26-Jan-63	21-PRES/63 DT 16 FEB 63
4	4492	FLT LT	BS KALRA	GD(P)	1965	26-Jan-65	20-PRES/65 DT 13 MAR 65
5	5391	FLT LT	R TANDON	GD(P)	1965	26-Jan-65	20-PRES/65 DT 13 MAR 65
6	4597	SQN LDR	BS BAKSHI	GD(P)	1966	1-Jan-66	14-PRES/66 DT 12 FEB 66
7	5707	FLT LT	JL DWELTZ	GD(P)	1966	1-Jan-66	14-PRES/66 DT 12 FEB 66
8	6351	FLT LT	SM HUNDIWALA	GD(P)	1966	1-Jan-66	14-PRES/66 DT 12 FEB 66
9	6506	FLT LT	LK DUTTA	GD(P)	1966	1-Jan-66	14-PRES/66 DT 12 FEB 66
10	6513	FLT LT	RK MALHOTRA	GD(P)	1966	1-Jan-66	14-PRES/66 DT 12 FEB 66
11	6532	FLT LT	CS KANWAR	GD(P)	1966	1-Jan-66	14-PRES/66 DT 12 FEB 66
12	30082	FLT LT	P GOSWAMI	GD(P)	1966	1-Jan-66	14-PRES/66 DT 12 FEB 66
13	208603	SGT	S SINGH	FLT ENG	1966	1-Jan-66	14-PRES/66 DT 12 FEB 66
14	4737	SQN LDR	RD PANT	GD(P)	1969	26-Jan-69	18-PRES/69 DT 12 APR 69
15	6756	FLT LT	MS SEKHON	GD(P)	1969	26-Jan-69	18-PRES/69 DT 12 APR 69
16	6891	FLT LT	HC DEMOS	GD(P)	1969	26-Jan-69	18-PRES/69 DT 12 APR 69
17	10384	FLT LT	MN SINGH	GD(P)	1969	26-Jan-69	18-PRES/69 DT 12 APR 69
18	4436	WG CDR	KK SAINI VrC	GD(P)	1970	26-Jan-70	16-Pres/70 DT 25 APR 70
19	5057	SQN LDR	N KUMAR	GD(P)	1970	26-Jan-70	16-Pres/70 DT 25 APR 70
20	9726	FLT LT	BS CHANDEL	F(P)	1972	26-Jan-72	28-PRES/73 DT 12 MAY 73
21	11366	FG OFFR	JP MATHUR	F(P)	1972	26-Jan-72	28-PRES/73 DT 12 MAY 73
22	7603	FLT LT	HS BEDI	F(P)	1972	23-Sep-72	108-PRES/72 DT 7 OCT 72
23	9412	FLT LT	PN SHARMA	F(P)	1972	23-Sep-72	108-PRES/72 DT 7 OCT 72
24	9524	FLT LT	RV SINGH	F(P)	1972	23-Sep-72	108-PRES/72 DT 7 OCT 72
25	9797	FLT LT	RT CHANDANI	F(P)	1972	23-Sep-72	108-PRES/72 DT 7 OCT 72
26	202634	WO	A THOMAS VM	FLT ENG	1973	26-Jan-73	34-PRES/74 DT 23 MAR 74
27	9535	FLT LT	T SINGH	F(P)	1974	26-Jan-74	85-PRES/74 DT 6 JUL 74
28	10954	FLT LT	PP CHOPRA	F(P)	1975	26-Jan-75	16-PRES/75 DT 21 FEB 76

Sl No	Ser No	Rank	Name	Branch/Trade	Year	Award Date	Gazette Notification Number
29	9821	FLT LT	HS AHLUWALIA	F(P)	1975	1-Jul-75	36-PRES/76 DT 1 MAY 76
30	10113	FLT LT	SC KHARBANDA	F(P)	1975	1-Jul-75	36-PRES/76 DT 1 MAY 76
31	11305	FLT LT	BLK REDDY	F(P)	1975	1-Jul-75	36-PRES/76 DT 1 MAY 76
32	11419	FLT LT	BR LOHTIA	F(P)	1975	1-Jul-75	36-PRES/76 DT 1 MAY 76
33	5694	SQN LDR	SR DESHPANDE	F(P)	1976	26-Jan-76	5-PRES/77 DT 22 JAN 77
34	5873	SQN LDR	KR DUTTON	F(P)	1976	26-Jan-76	5-PRES/77 DT 22 JAN 77
35	8441	SQN LDR	VK MOHOTRA	F(P)	1977	26-Jan-77	20-PRES/78 DT 27 MAY 78
36	10593	FLT LT	TR SINGH	F(P)	1978	26-Jan-78	67-PRES/79 DT 22 DEC 79
37	7019	WG CDR	K SRIDHARAN	F(P)	1982	26-Jan-82	1-PRES/83 DT 22 JAN 83
38	8424	WG CDR	A SINGH	F(P)	1984	20-Mar-84	31-PRES/84 DT 7 APR 84
39	11290	SQN LDR	TPS CHHATWAL	F(P)	1984	20-Mar-84	31-PRES/84 DT 7 APR 84
40	10502	SQN LDR	AJ RAO	F(P)	1985	26-Jan-85	29-PRES/85 DT 6 APR 85
41	10514	SQN LDR	GMS BAJWA	F(P)	1985	26-Jan-85	29-PRES/85 DT 6 APR 85
42	11288	SQN LDR	SS BAINS	F(P)	1985	26-Jan-85	29-PRES/85 DT 6 APR 85
43	10987	SQN LDR	SC SOLOMAN	F(P)	1986	26-Jan-86	35-PRES/86 DT 19 APR 96
44	12008	SQN LDR	HS BATH	F(P)	1986	26-Jan-86	35-PRES/86 DT 19 APR 96
45	11276	WG CDR	MR HANDA	F(P)	1987	26-Jan-87	37-PRES/87 DT 25 APR 87
46	11433	WG CDR	IM SIMOES	F(P)	1987	26-Jan-87	37-PRES/87 DT 25 APR 87
47	15875	FLT LT	ANSHU KUMAR MATTA	F(P)	1987	26-Jan-87	37-PRES/87 DT 25 APR 87
48	263186	JWO	B SINGH	FLT GNR	1987	26-Jan-87	37-PRES/87 DT 25 APR 87
49	11300	WG CDR	JS GUJRAL	F(P)	1988	26-Jan-88	24-PRES/88 DT 2 APR 88
50	11579	WG CDR	SC MALHAN	F(P)	1988	15-Apr-88	78-PRES/88 DT 13 AUG 88
51	13165	SQN LDR	VSN NATH	F(P)	1988	15-Apr-88	78-PRES/88 DT 13 AUG 88
52	13398	SQN LDR	AD SONPAR	F(P)	1988	15-Apr-88	78-PRES/88 DT 13 AUG 88
53	13946	SQN LDR	RS KATARIA	F(P)	1988	15-Apr-88	78-PRES/88 DT 13 AUG 88
54	16213	FLT LT	S BISHNOI	F(P)	1988	15-Apr-88	78-PRES/88 DT 13 AUG 88
55	16587	FLT LT	R BALKRISHAN	F(P)	1988	15-Apr-88	78-PRES/88 DT 13 AUG 88
56	11663	WG CDR	HS SODHI	F(P)	1989	26-Jan-89	39-PRES/89 DT 22 APR 89
57	18257	FG OFFR	MR ANAND	F(P)	1989	26-Jan-89	39-PRES/89 DT 22 APR 89
58	641079	SGT	B SINGH	ENG FIT	1989	26-Jan-89	39-PRES/89 DT 22 APR 89
59	658428	CPL	T VENU MADHAVAN	AF FIT	1989	26-Jan-89	39-PRES/89 DT 22 APR 89

Sl No	Ser No	Rank	Name	Branch/Trade	Year	Award Date	Gazette Notification Number
60	11312	WG CDR	AK SAXENA	F(P)	1990	26-Jan-90	38-PRES/90 DT 12 MAY 90
61	13812	SQN LDR	AS SLAICH	F(P)	1990	26-Jan-90	38-PRES/90 DT 12 MAY 90
62	15220	SQN LDR	KULWANT SINGH GILL	F(P)	1990	26-Jan-90	38-PRES/90 DT 12 MAY 90
63	17145	FLT LT	A DOGRA	F(P)	1990	26-Jan-90	38-PRES/90 DT 12 MAY 90
64	17336	FLT LT	HS AUJLA	F(P)	1990	26-Jan-90	38-PRES/90 DT 12 MAY 90
65	14281	SQN LDR	HPS NATT	F(P)	1991	26-Jan-91	3-PRES/91 DT 24 AUG 91
66	14295	SQN LDR	KNG NAIR	F(P)	1991	26-Jan-91	3-PRES/91 DT 24 AUG 91
67	16549	SQN LDR	SG WARRIER	AE(M)	1991	26-Jan-91	3-PRES/91 DT 24 AUG 91
68	19181	FG OFFR	S SOOD	F(P)	1991	26-Jan-91	3-PRES/91 DT 24 AUG 91
69	228616	MWO	DS BINDRA	FLT ENG	1991	26-Jan-91	3-PRES/91 DT 24 AUG 91
70	659174	SGT	Y KALLURI	ENG FIT	1991	26-Jan-91	3-PRES/91 DT 24 AUG 91
71	11442	GP CAPT	FH MAJOR	F(P)	1992	26-Jan-92	29-PRES/92 DT 11 APR 92
72	16633	SQN LDR	VVS SUBBARAO	F(P)	1992	26-Jan-92	29-PRES/92 DT 11 APR 92
73	12027	WG CDR	M RAMAKRISHNA	F(P)	1993	26-Jan-93	63-PRES/93 DT 24 APR 93
74	18556	FLT LT	RK CHAUHAN	F(P)	1993	26-Jan-93	63-PRES/93 DT 24 APR 93
75	18581	FLT LT	P UPADHYAY	F(P)	1993	26-Jan-93	63-PRES/93 DT 24 APR 93
76	17471	SQN LDR	AP SRIVASTAVA YSM	F(P)	1999	15-Aug-99	27-PRES/2000 DT 11 MAR 2000
77	18298	SQN LDR	NS VERMA	F(P)	1999	15-Aug-99	27-PRES/2000 DT 11 MAR 2000
78	19898	SQN LDR	N KUMAR	F(P)	1999	15-Aug-99	27-PRES/2000 DT 11 MAR 2000
79	22739	FLT LT	S MUHILAN (POSTHUMOUS)	F(P)	1999	15-Aug-99	27-PRES/2000 DT 11 MAR 2000
80	668396	WO	KS DHILLON	FLT ENGR	1999	15-Aug-99	27-PRES/2000 DT 11 MAR 2000
81	695490	SGT	PPVN RAVI (POSTHUMOUS)	FLT GNR	1999	15-Aug-99	27-PRES/2000 DT 11 MAR 2000
82	699103	SGT	T GHOSH	FLT ENGR	1999	15-Aug-99	32-PRES/2000 DT 11 MAR 2000
83	729917	SGT	SR KISHORE (POSTHUMOUS)	FLT ENGR	1999	15-Aug-99	27-PRES/2000 DT 11 MAR 2000
84	17347	SQN LDR	P VISHNOI	F(P)	2000	26-Jan-00	85-PRES/2000 DT 26 JAN 2000
85	17706	SQN LDR	PE PATANGE YSM	F(P)	2000	15-Aug-00	150-PRES/2000 DT 30 DEC 2000
86	21573	SQN LDR	K GURAO	F(P)	2000	15-Aug-00	150-PRES/2000 DT 30 DEC 2000
87	22374	FLT LT	RAJAT SAHA	AE(M)	2000	15-Aug-00	150-PRES/2000 DT 30 DEC 2000
88	15869	WG CDR	RK NEGI	F(P)	2001	26-Jan-01	68-PRES/2001 DT 26 MAY 01
89	17002	WG CDR	AT SAMTANI	F(P)	2001	26-Jan-01	68-PRES/2001 DT 26 MAY 01
90	17722	SQN LDR	G THOMAS	F(P)	2001	26-Jan-01	68-PRES/2001 DT 26 MAY 01

Sl No	Ser No	Rank	Name	Branch/Trade	Year	Award Date	Gazette Notification Number
91	18580	SQN LDR	D SINGH	F(P)	2001	26-Jan-01	68-PRES/2001 DT 26 MAY 01
92	22726	FLT LT	P DIXIT	F(P)	2001	26-Jan-01	68-PRES/2001 DT 26 MAY 01
93	16968	WG CDR	RAJESH ISSER	F(P)	2001	15-Aug-01	123-PRES/2001 DT 19 JAN 02
94	18303	SQN LDR	GS PADDA	F(P)	2001	15-Aug-01	123-PRES/2001 DT 19 JAN 02
95	20489	SQN LDR	S SHARMA	F(P)	2001	15-Aug-01	123-PRES/2001 DT 19 JAN 02
96	21015	SQN LDR	A SHARMA	F(P)	2001	15-Aug-01	123-PRES/2001 DT 19 JAN 02
97	22982	FLT LT	AS GAHARWAR	F(P)	2001	15-Aug-01	123-PRES/2001 DT 19 JAN 02
98	23746	FLT LT	RC PATHAK	F(P)	2001	15-Aug-01	123-PRES/2001 DT 19 JAN 02
99	641288	WO	KC DAS	FLT GNR	2001	15-Aug-01	123-PRES/2001 DT 19 JAN 02
100	679540	JWO	SK SINGH	FLT GNR	2001	15-Aug-01	123-PRES/2001 DT 19 JAN 02
101	20155	SQN LDR	SP JOHN	F(P)	2002	26-Jan-02	118-PRES/2002 DT 29 JUN 02
102	18838	SQN LDR	SK BHATNAGAR	F(P)	2004	26-Jan-04	32-PRES/2004 DT 24 APR 04
103	22908	SQN LDR	N BHALERAO	F(P)	2004	26-Jan-04	32-PRES/2004 DT 24 APR 04
104	17453	WG CDR	MUKESH RAWAT	F(P)	2004	15-Aug-04	131-PRES/2004 DT 13 NOV 04
105	19037	SQN LDR	RAKSHALE SANJEEV KUMAR	AE(L)	2004	15-Aug-04	131-PRES/2004 DT 13 NOV 04
106	24035	FLT LT	NAVIN MARANATHA NAYAR	F(P)	2004	15-Aug-04	131-PRES/2004 DT 13 NOV 04
107	24229	FLT LT	MAYANGALAMBAM MANIMOHON SINGH	F(P)	2004	15-Aug-04	131-PRES/2004 DT 13 NOV 04
108	19570	WG CDR	SHASHI KANT MISHRA	F(P)	2005	26-Jan-05	28-PRES/2005 DT 9 APR 05
109	19896	SQN LDR	MANU CHAUDHARY	F(P)	2005	26-Jan-05	28-PRES/2005 DT 9 APR 05
110	22345	SQN LDR	RANDHIR SINGH	F(P)	2005	26-Jan-05	28-PRES/2005 DT 9 APR 05
111	17323	GP CAPT	UMESH KUMAR SHARMA	F(P)	2005	15-Aug-05	130-PRES/2005 DT 26 NOV 05
112	18101	WG CDR	SUBRAMANIAM RAVI VRIDHACHALEM	F(P)	2005	15-Aug-05	130-PRES/2005 DT 26 NOV 05
113	18831	WG CDR	BHANU JOHRI	F(P)	2005	15-Aug-05	130-PRES/2005 DT 26 NOV 05
114	20163	WG CDR	MANISH GIRIDHAR	F(P)	2005	15-Aug-05	130-PRES/2005 DT 26 NOV 05
115	20521	WG CDR	ASHOK PANDE	F(P)	2005	15-Aug-05	130-PRES/2005 DT 26 NOV 05
116	25338	FLT LT	DEVAVRAT JAGDISH BHANDARKAR	F(P)	2005	15-Aug-05	130-PRES/2005 DT 26 NOV 05
117	765876	SGT	DAVENDRA KUMAR SHARMA	FLT GNR	2005	15-Aug-05	130-PRES/2005 DT 26 NOV 05
118	22901	SQN LDR	GS BHULLAR	F(P)	2006	15-Aug-06	104-PRES/2006 DT 15 AUG 06

Sl No	Ser No	Rank	Name	Branch/ Trade	Year	Award Date	Gazette Notification Number
119	20781	WG CDR	SUBROTO KUNDU	F(P)	2007	15-Aug-07	168-PRES/2007 DT 17 NOV 07
120	21009	WG CDR	JITENDER YADAV	F(P)	2007	15-Aug-07	168-PRES/2007 DT 17 NOV 07
121	21024	WG CDR	SARTAJ BEDI	F(P)	2007	15-Aug-07	168-PRES/2007 DT 17 NOV 07
122	21812	WG CDR	RAJESH RANWAN	F(P)	2007	15-Aug-07	168-PRES/2007 DT 17 NOV 07
123	24203	SQN LDR	BHUPENDER SINGH SEHRAWAT	F(P)	2007	15-Aug-07	168-PRES/2007 DT 17 NOV 07
124	768541	CPL	PRASANTA KUMAR PATI	ENG FIT	2007	15-Aug-07	168-PRES/2007 DT 17 NOV 07
125	18793	WG CDR	AJAY SHUKLA	F(P)	2008	26-Jan-08	34-PRES/2008 DT 10 MAY 08
126	26715	FLT LT	AMIT SHARMA (POSTHUMOUS)	F(P)	2008	26-Jan-08	34-PRES/2008 DT 10 MAY 08
127	20447	WG CDR	HARVINDER SANDHU	F(P)	2008	15-Aug-08	102-PRES DT 08 NOV 08
128	21015	WG CDR	A SHARMA VM	F(P)	2009	26-Jan-09	21-PRES/2009 DT 21 MAR 09
129	22545	WG CDR	DK VATS	F(P)	2009	26-Jan-09	22-PRES/2009 DT 21 MAR 09
130	658793	WO	PS RAJPUT	FLT GUN	2009	26-Jan-09	22-PRES/2009 DT 21 MAR 09
131	729086	JWO	JH SIDDIQI	FLT ENG	2009	26-Jan-09	22-PRES/2009 DT 21 MAR 09
132	21846	WG CDR	RAJAN PURI	F(P)	2009	15-Aug-09	118-PRES/2009 DT 05 DEC 09
133	28497	FLT LT	YS TOMAR	F(P)	2009	15-Aug-09	118-PRES/2009 DT 05 DEC 09
134	679560	JWO	RAJHANS THAPLIYAL	FLT GUN	2009	15-Aug-09	118-PRES/2009 DT 05 DEC 09
135	22711	WG CDR	SAGAR SINGH RAWAT	F(P)	2010	15-Aug-10	115-PRES/2010 DT 23 OCT 10
136	24900	SQN LDR	ASHOK KUMAR	F(P)	2010	15-Aug-10	115-PRES/2010 DT 23 OCT 10
137	27475	SQN LDR	VISHAL MOHAN	F(P)	2010	15-Aug-10	115-PRES/2010 DT 23 OCT 10
138	21578	WG CDR	VIVEK AHLUWALIA	F(P)	2011	15-Aug-11	117-PRES/2011 DT 05 NOV 11
139	29736	FLT LT	HASHEER HAMEED	F(P)	2011	15-Aug-11	117-PRES/2011 DT 05 NOV 11
140	23193	WG CDR	N WELDE	F(P)	2012	26-Jan-12	24-PRES/2012 DT 24 MAR 12
141	24221	WG CDR	M SHYLU	F(P)	2012	26-Jan-12	24-PRES/2012 DT 24 MAR 12

Bibliography

- Interviews with many retired officers and Commanding Officers of IAF Helicopter Units and Unit Diaries & Records.
- Gallantry Award Citations; Govt of India Gazzettes & Air HQ Records.
- "My years with the IAF"; Air Chief Marshal PC Lal.
- "The Indian Air Force Memorial Book"; Sqn Ldr RTS Chinna (retd), N Delhi, Air HQ 1996.
- "For the Honour of India" A History of Peacekeeping of the Indian Armed Forces; Gen Satish Nambiar (retd) & Team, CAFHR, USI, N Delhi.
- History of the Indo-Pak War 1965 (MoD).
- History of the Indo-Pak War 1971 (MoD).
- http://www.airhq.iaf.in.
- Website of Bharat Rakshak.
- "Lessons of Operations in Afghanistan"; Col Nikolay Spasibo, Conflict Studies research Centre (CSRC), UK 1997.
- "The Bear went over the Mountain"; Lester E Grau (ed) NDU Washington DC 1996.
- "The Use of Russian Air Power in Second Chechen War"; Marcel de Haas, B59, CSRC 2003.
- "High Altitude Warfare: The Kargil Conflict and the Future", Marcus de Costa Jun 2003, NPS Monterey.
- "Learning Large Lessons: The Evolving Roles of Ground Power and Air Power in the Post Cold War Era", David E Johnson, RAND 2007.
- The USAF RPV & UAV Strategic Vision 2005, USAF.
- FM 1-112: Attack Helicopter Operations, HQ Dept of the Army, Washington 1997.
- "North Caucasus: Problems of Helicopters Support in Mountains", CN Blandy Aug 2007, CSRC 07/24.

WE DO THE DIFFICULT AS A ROUTINE
THE IMPOSSIBLE MAY TAKE A BIT LONGER

- "Ground Combat at High Altitude", Lester E Grau, Military Review Bulletin, US.
- "Mountain Warfare is not the only thing slowing down the US Army", Jason Vest, Defence & National Interest Website, Mar 2002.
- "Command & Control of Battlefield Helicopter: The Search for a Joint Approach", Martin Sharp, Air Power Studies Centre, Canberra, 1998.
- "Mujahideen Tactics in the Soviet Afghan War", CJ Dick, CSRC Jan 2002.
- "Evaluating Russian Performance in the second Chechen War", Michael Orr, JDW Mar 2000.
- "Air Power Against Terror: America's conduct of OEF", Benjamin S Lambeth, RAND Corporation, 2005.
- "Air Power in land Operations", Gurmeet Kanwal, Air Power Journal Vol 2, No. 1 Spring 2005, CAPS New Delhi.
- "Attack helicopters"; Peter Donaldson; and, "Electronic Warfare"; David S Harvey Defence Helicopter (May/Jun 07).
- "German Army Orders"; Thomas Newdick, Defence Helicopter– Aug 2006.
- "RPG Anti-aircraft on the Cheap"; Col David Eshel, Defence Helicopter– December 2005/Jan 2006.
- "On Target for New Missions"; Tom Withington, Defence Helicopter– Nov/Dec 2007.
- "USAF Lines up IFFCC Work"; Gareth Jennings and Caitlin Harrington; Jane's International Defence Review-Dec 07 edition.
- "Attack Helicopters Adapt their Role for the Asymmetric Battlefield"; Rupert Pengelly, Jane's International Defence Review-Dec 07 edition.
- "Front-line Helicopter Gunships Evolve for the Modern Battlefield"; Michael J Gething, Jane's International Defence Review-Dec 07 edition.
- "Inside the Soviet Army in Afghanistan"; Alexiev A, RAND Corporation, Santa Monica 1988.
- "The War in Chechnya"; GD Bakshi, A Military Analysis; Strategic Analysis IDSA, Aug 2000 (Vol xxiv No 5).
- FM 1-100, Doctrinal Principles for Army Aviation in Combat Operations; US Army
- "The Afghan-Pakistan War: Casualties, the Air War, and Win, Hold, Build"; Anthony H. Cordesman, Arleigh A. Burke Chair in Strategy,CSIS May 15, 2009

- "Air Combat Trends in the Afghan and Iraq Wars"; Anthony H. Cordesman, Arleigh. Burke Chair in Strategy; CSIS, March 11, 2008.

- "Preparing and Training for the Full Spectrum of Military Challenges; Insights from the Experiences of China, France, the United Kingdom, India, and Israel". Johnson, Moroney, Cliff, Smallman, Spirtas; RAND 2009.

- "Posture Statement of Admiral Eric Olson, USN Commander, United States Special Operations Command Before the 112th Congress senate Armed Services Committee, March 1, 2011.

- "Special Operations Forces Aviation at the Crossroads"; Clark A. Murdock, Rebecca Grant, Richard Comer' Thomas P. Ehrhard; CSIS, 2007.

- "Airpower and Counterinsurgency: Building on a Proper Foundation"; Paul Smyth, http://www.airpower.au.af.mil.

- "The Development of China's Air Force Capabilities"; Roger Cliff, CT-346 May 2010; Testimony presented before the U.S.-China Economic and Security Review Commission on May 20, 2010.